The APE in the CORNER
OFFICE

Also by Richard Conniff

The Natural History of the Rich: A Field Guide

Every Creeping Thing: True Tales of Faintly Repulsive Wildlife

Spineless Wonders: Strange Tales from the Invertebrate World

The Devil's Book of Verse: Masters of the Poison Pen from Ancient Times to the Present Day

The APE in the CORNER OFFICE

Understanding the Workplace Beast in All of Us

Richard Conniff

CROWN BUSINESS
NEW YORK

Grateful acknowledgment is made to Paul Ekman for facial expression photos from his training CDs, available at www.paulekman.com.

Published in the United States by Crown Business, an imprint of the Crown Publishing Group, a division of Random House, Inc., New York.
www.crownpublishing.com

Crown Business is a trademark and the Rising Sun colophon is a registered trademark of Random House, Inc.

Library of Congress Cataloging-in-Publication Data

Conniff, Richard, 1951–
The ape in the corner office : understanding the workplace beast in all of us / Richard Conniff.—1st ed.
1. Office politics. 2. Interpersonal relations. 3. Supervision of employees. 4. Power (Social sciences). 5. Corporate culture. I. Title.
HF5386.5.C66 2005
650.1′3—dc22 2005006407

ISBN 1-4000-5219-X

Printed in the United States of America

Design by Lauren Dong

10 9 8 7 6 5 4 3 2 1

First Edition

For cubicle monkeys everywhere,
in the hope that this book
will help them lift up their heads—
and hoot.

CONTENTS

The APE in the CORNER
OFFICE

1

YES, IT IS A GODDAMN JUNGLE OUT THERE
Why Acting Like an Animal Comes So Easy

Animals in the wild lead lives of compulsion and necessity within an unforgiving social hierarchy in an environment where the supply of fear is high and the supply of food low and where territory must constantly be defended and parasites forever endured.

—YANN MARTEL, *Life of Pi*

Sounds like an average day at the office, doesn't it? Compulsion, necessity, the unforgiving social hierarchy, parasites . . . Oh, and the high supply of fear. That one I could feel butterfly-fluttering in my abdomen and ant-dancing out on the fringes of my peripheral nervous system. I was standing in front of the top North American distributors for a leading European manufacturer. We had assembled at a resort in the Grand Tetons, in an area still populated by grizzly bears and gray wolves, to which I expected shortly to be thrown. I'd been asked to give a talk about how businesspeople act like animals. I was vaguely nervous.

The top baboon for the North American division, a big, bluff fellow, sat in the front row, arms folded, with his wife (blond, witty, appealing) to one side and his head of sales (short, round, ebullient) on the other. At dinner the night before I had gotten to know many of these people by first name. I recalled a quote about how businesspeople "don't like being compared to bare-ass monkeys." I took a deep breath.

Everybody in the room had heard the statistic that humans

are roughly 99 percent genetically identical to chimpanzees. By some estimates, the difference between our two species may be a matter of fewer than fifty genes, out of perhaps twenty-five thousand shared in common. But hardly anyone in the business world seems to have considered what that might mean in our working lives. More often than not, managers endeavor to minimize the human, much less the animal, element and make companies hum like machines. In their own lives, individual workers also tend to treat human nature mainly as something to be overcome, by getting the hair waxed from their torsos or added to their scalps, by dressing for success, by giving at least the appearance of handling stress. (Was that the serene brow of Botox I detected on a woman in the first row? It was really too early in my talk for her to be numb with boredom.)

I asked my audience to think for a moment about how their everyday workplace behavior might be shaped by forces that are less susceptible to change—by the drives and predispositions bequeathed to us by our long evolution first as animals and later as tribal humans. By fear. By anger. By the primordial yearning for social allies and for status. Think of yourself, I suggested, as part of a primate hierarchy unconsciously following thirty-million-year-old rules for establishing dominance and submission, for waging combat and maintaining peace. Think about how the alpha, whether chimpanzee or chief executive officer, typically asserts authority with the identical language of posture, stride, lift of chin, directness of gaze, the sharp glower to quell an unruly subordinate.

The head guy in the first row started to light up at this, especially when I got to the stuff about using political maneuvering among chimpanzees as a better way to understand boardroom confrontations. He surged out of his seat when the talk was done and launched into what he called the natural history of the boardroom.

In the upper echelons at company headquarters, he said, the conference tables are circular rather than rectangular, ostensibly for a roundtable atmosphere of equality. "Well, *bollocks,*" he said. In fact, there is a distinct hierarchy, and everybody knows where everybody else stands, or sits, in it; the circular form merely makes the combat a little more open. In a week or two, he said, he'd be heading overseas for a meeting of a committee where the chairman had lately vacated his seat. "No one will say anything. But everyone will be looking at that seat and wondering who's going to take it, whether anyone will have the audacity to sit there."

"You should sit there," the head of sales ventured.

"No, I'd be like the baboon trying to rise three steps above his rank—I'd get knocked down." He was a realist, yet keen for the combativeness that would inevitably surface. "I *love* it," he said. "Sometimes when there's a kill about to happen, there's a moment of hesitation when people aren't sure if it's going to happen."

By now my eyes were beginning to widen.

"And then they get the scent, and they know it's going to be okay, and they know who's going to take the lead, and who's going to come in for the kill."

"It's like the Serengeti," the sales guy agreed. "The round table just makes it easier for everybody to see the kill."

"Jesus," I said.

"Don't worry," the head guy's wife interjected, taking him gently by the elbow. *"I'm* really in control here." And everybody laughed.

THIS COMPANY IS A ZOO

Maybe I shouldn't have been surprised that some businesspeople are in fact entirely prepared to liken themselves to bare-ass monkeys. They just want to be *dominant, predatory* bare-ass monkeys. Animal analogies have always ranked among the favorite clichés of the business world, where eight-hundred-pound gorillas run with the big dogs, swim with the sharks, occasionally find themselves up to their asses in alligators, and, if they are not crazy like a fox, can end up caught like a deer in the headlights.

When Richard Kinder quit Enron to form his own gas company in 1996, he disguised his dismay with Kenneth Lay's leadership under a standard animalism: "If you aren't the lead dog, the scenery never changes." H. Ross Perot also resorted to animal analogies when he was tormenting the hapless, imperial General Motors CEO Roger Smith:

"Revitalizing General Motors is like teaching an elephant to tap-dance. You find the sensitive spots and start poking." (Or did he say "lap dance"? In any case, Lou Gerstner at IBM knew a good line when he saw it, and stole it for the title of his book, *Who Says Elephants Can't Dance?*) Even the eminently clever satirist Scott Adams ended up likening almost everybody in the working world of his antihero Dilbert to a weasel.

The truth beneath the clichés is that the lives of animals are not nearly so simple as we used to think. Nor are the lives of working people so complex as we like to believe. Moreover, the two have a lot in common, and not just in the obvious ways. For instance, aggressive business types often employ animal analogies because they mistake them for *The Art of War* by other means. The idea of animal troops ruled by "demonic males" dishing out "nature, red in tooth and claw" appeals to a certain view of business life: *It really is a goddamn jungle out there.* And don't get me wrong. This is a very entertaining view, and one I intend to indulge fully over the course of this book. Like my North American division chief, we all love a good brawl, if only from a safe distance.

But it's also a narrow, misleading point of view. Here's the sort of surprising thing we can learn from a more careful look at the animal world: Even chimps spend only about 5 percent of their day in aggressive encounters. By contrast, they devote as much as 20 percent of the working day to grooming family, friends, and even subordinates. When they fight with rivals in the troop, they often go well out of their way, after the dust settles, to kiss and make up. And why should working people care how chimpanzees resolve their conflicts? Because our social behaviors and theirs evolved from the same ancestors and still follow many of the same rules. In one case described later in this book, a better understanding of the nature of reconciliation saved a company $75 million in litigation and insurance costs. Even in our everyday working lives, human bosses, like alpha chimps, sometimes drive their underlings beyond any reasonable limits. They might do better in life (and in business) if they understood just how far even a dumb ape will go to achieve harmony in the aftermath of conflict.

BITE THAT METAPHOR

Businesspeople regularly trot out animal analogies that make no sense. Despite their reputation as cold-eyed realists, they apparently have trouble separating fact from ridiculous fiction. You can do better:

Ostriches don't bury their heads in the sand. In fact, ostriches merely lower their heads to the ground to avoid detection while keeping an eye out for danger. Some biologists suggest that they are trying to disguise the 400-pound bulk of their torsos as a termite mound. But in the African savanna where they live, actually burying one's head in the sand would be a good way to get bitten on the ass by a lion. (What biologists call "nonadaptive behavior.")

Lemmings don't leap off cliffs to commit mass suicide. When a population boom causes overcrowding, these Arctic rodents do the sensible thing and migrate en masse in search of a new home. A few of them may occasionally get crowded off a ledge as they swarm into unfamiliar territory. But it's an accident. Really. The myth of mass suicide got enshrined in modern urban lore by Disney filmmakers in the 1950s, who had the dumb idea that forcing captive lemmings off a cliff would make for dramatic film footage.

Real weasels don't wear tassels on their shoes. And they spend most of their time chasing down mice, rats, and other rodents. This makes them heroes, not villains, contrary to the chickenhouse myth. So if it's not right to call your typical slimy record industry executive a "weasel," what should you call him? Just say "sleazebag" and leave innocent animals out of it.

Sales reps sometimes talk about having a "salmon day:" You spend the whole day swimming upstream only to get screwed and die. But the sad truth is that salmon don't even get screwed. They merely spill their seed and their eggs on the streambed, leaving the little gametes to mix it up on their own. Meanwhile, the happy couple goes belly up and drifts away to get eaten by a bear.

Then again, if you like animal analogies, this is probably a more realistic characterization of a day in the life of the average salesperson.

EMOTIONAL ANIMALS

Shining an evolutionary light on the workplace isn't just a clever way to rationalize bad behavior, or to find simple-minded justifications for maintaining the status quo. It is a useful approach to survival in the workplace. Moreover, it's an approach that applies to any workplace, whether the workers happen to be greeting customers at a Wal-Mart in Los Angeles, or hanging Warhols at the Tate Modern in London, or stamping out toasters at a Haier Company plant in Qingdao, China. Understanding evolutionary propensities can help us manage conflict, build useful alliances, avoid backstabbings, survive boardroom assassination attempts, and understand the unspoken emotions revealed by the facial expressions of the people around us.

It can at times help companies manage the workplace to accommodate things people do naturally. For instance, W. L. Gore & Co., maker of Gore-Tex, has chosen to keep its plants and offices at what feels like a comfortable human scale—under two hundred employees. This is close to the maximum size of the tribal clans in which human society evolved, and some biologists say our brains are actually built to operate on this social scale. It isn't what most companies mean when they talk about right-sizing. But Gore employees say it feels like the right size for working together effectively.

Ignoring evolutionary and biological propensities, on the other hand, often proves disastrous. For instance, the U.S. military has traditionally organized itself into companies of roughly the same tribal clan size, subdivided into platoons where thirty or so soldiers train together and develop the kind of tight, cohesive bond needed for deadly combat. But in the 1960s corporate-style managers tried to rethink this traditional structure and override human nature. Managerial types have always relished the idea of business as war. But it was a much worse mistake to believe they could wage war as business.

The introduction of assembly line practices meant that soldiers rotated into the Vietnam War as individuals on twelve-month tours, not as part of a tightly bonded social group. Their officers moved through even faster, getting their combat tickets punched for the purpose of career advancement, instead of having their lives bound to the survival and success of their foot soldiers. One victim of this thinking later suggested that the military would have been better off had it treated its troops literally like dogs: "Now, this is strange: The only reason I got to go to 'Nam with the unit I trained with is because I was in the Canine Corps. See, the army knew that the dogs would get depressed if they were broken up, so they kept them together. So the trainers got to stick together, too. But almost everybody else went in alone. Just interchangeable parts, like ammo clips or mortar shells. The army figured if you were a mortar man, you could do your job in any unit, so it didn't matter what unit you were with or if you had any buddies there or anything. It's funny when you think about it. Funny and sick. The army knew it was bad for the dogs to get split up, but it was okay for guys to go to hell all alone."

This is the characteristic error of our time, if not of our species: We tell ourselves that we are rational beings, not animals, that we are in control of our postbiological world. We certainly don't allow biology or emotions to control us. "I don't do feelings," Sun Microsystems CEO Scott McNealy recently declared, with apparent contempt. "I'll leave that to Barry Manilow." In this view, work (and even warfare) is a purely logical business of balancing pros and cons to maximize profit and minimize loss. The truth, of course, is the opposite: We are emotional animals, and an evolutionary and anthropological view of the workplace is essential for survival in a competitive struggle that seems increasingly Darwinian.

The workplace has supplanted the tribe, the community, and even the family as the focus of our lives, and it has become the arena for all the behaviors that originally evolved in those contexts. Understanding our evolutionary and anthropological propensities has become even more important as our jobs have become less secure. The realization has crept up the ranks from the assembly line to customer service to engineers and even upper management that their jobs could be outsourced any day now to someone in Bangalore willing to do the same work for a fraction of the pay. (Lehman Brothers and Bear Stearns now hire MBA financial analysts in India starting at $800 a month.) Workers in Bangalore are in turn acutely aware that a shift in currency rates or a period of political turmoil could just as easily send their jobs to Kuala Lumpur or to Ciudad Juarez.

Even if we manage to keep our jobs, we now routinely participate via high technology in work groups with people in such distant locales. It's becoming a job requirement to get past mere cultural differences and figure out how to replicate ties of trust, comfort, collaboration, and hierarchy that existed, only two or three generations ago, in the family, or the platoon, or the neighborhood. And we need to make these ties work through virtual media and across oceans. Understanding the nature and origin of human behavior is the only way to do this job well.

CORPORATE APES

Talking about the biological basis of our behaviors is likely to irritate some people. We may joke about eight-hundred-pound gorillas, but we also like to believe we have moved confidently into the technological era and put our old biological constraints behind us. We spend our days in front of computer screens, multitasking in the e-zone. "Our bodies were abandoned long ago, reduced to hunger and sleeplessness and the ravages of sitting for hours at a keyboard and a mouse," writes Silicon Valley computer programmer Ellen Ullman in her 1997 book *Close to the Machine*. "Our physical selves had been battered away. Now we know each other in one way and one way only: the code."

In some jobs, actually seeing colleagues has become so rare we have a buzzword for it: "face time." At meetings, we schedule "bio-breaks," as if grabbing a bite or going to the bathroom were the only things remotely animal-like about our lives. One U.S. insurance company, now in leaner, meaner mode, even frowns on grabbing a bite. In the old days, a former executive there says, the guy who had to leave a meeting first to go to the bathroom was definitely "the B-dog." Now, the company schedules back-to-back meetings into the night and refuses to pay for meals for people working late. The unspoken culture also treats bringing a snack to a meeting as a mark of weakness, the executive says. "Needing food has become the new going to the bathroom."

Meanwhile, our knees, our bellies, our animal hearts are quaking.

We've had at best a century or two in the business of being company employees. Until the Industrial Revolution of the 1820s, about 80 percent of the workforce was self-employed. That is, they lived much as humans always had, working with a few friends and relatives, in small communities, following the shape of the seasons, always keeping an eye out for the main chance. By the time William H. Whyte Jr.'s *The Organization Man* appeared in 1956, self-employment had dropped to a mere 18 percent of the workforce in the developed world. It stands at just 10 percent today, even taking into account the supposedly liberating effects of the Internet, the cell phone, and the telecommute.

We pass our workdays in a context neither our human nor simian forebears would consider natural. Or rather, our time in the workplace is about as natural as a chimpanzee's in the zoo. (At some companies, it's as natural as a hen in a battery cage half pecked to death by some feathered tyrant.) The only thing that remains the same is the animal within.

Ullman, the computer programmer, is only fooling when she indulges the pleasant myth of our working selves as pure, disembodied intellect. She knows the underlying animal is all too present, as when she peels off her rain-soaked outer garments and sits down to join a fellow programmer at a sushi joint: "His head was thrown back, his eyes were half closed. He sniffed at the air—once, twice, three times. Then he actually snorted. Though I'm sure I've never seen anything like it in the whole of my life, I only needed to be a primate to understand its meaning: Brian wanted to fuck me." Millions of years of evolution have made us apes first of all and hunter-gatherers second, and, oh, Lord, how we carry those ancient behavioral patterns with us into the workplace.

Modern corporate citizens are of course much more than apes. We have free will. Nature does not, in fact, force us to behave one way or another. We do not automatically become what our genes might seem to dictate. Even in rats specifically bred to display a high level of anxiety, the nurturing behavior of a good mother can dramatically alter the expression of the genes, making her offspring better adjusted and less anxiety-prone in later life. And if a healthy environment can help anxious rats, surely there is hope for middle managers, too?

At the same time, it is naive to continue to ignore the many ways genetic propensities influence our behavior. For a given trait—aggressive inclinations, say, or a tendency to depression—scientists believe an individual's genetic heritage may account for up to half of his or her behavior. Workplace behaviors that we take to be no more than the whim of the moment often turn out, on closer examination, to be rooted millions of years deep in our biology. Understanding these roots can be a revelation.

THE EVOLUTION OF A SMILE

Take something as simple as a smile.

At a firm in New York City, employees were puzzled by a junior staffer named Chip, who greeted them, one coworker reports, "with an enthusiastic grin accompanied by a shrug of the shoulders, a slight self-deprecating cock of the head down and forward, and a wave of the hand that resembles someone swearing an oath in court." Chip repeated the grin and wave "every time he saw you. Every time he passed your office, every time you passed his," the seventh greeting of the day being as heartfelt as the first. It was disconcerting, and probably should have been. There is no biological signal more reassuring than the right smile in the right place, and few things so worrisome as when somebody gets it wrong.

Smiling is our oldest and most natural expression, and like other facial expressions, it evolved for a function, as a means of responding to the people around us and influencing their behavior. Primatologists connect our smile to the "fear grin" in monkeys and date its evolution back at least thirty million years. In a group of macaques, for example, the approach of the alpha may cause a subordinate to cringe and nervously pull back the corners of the mouth, exposing the clenched teeth. It's a signal meaning, "I'm no threat."

For humans, too—and not just Chip—this sort of smile is a way to disarm and reassure those around us, particularly our social superiors. And apparently the quicker the better: Threat reduction is so important to our survival that this function is deeply embedded in our facial physiology. Our smile muscles are 90 percent fast-twitch fibers, built for rapid response. By contrast, the muscle that furrows the brow—what Charles Darwin called "the muscle of difficulty"—lags behind at just 50 percent fast-twitch. Out on the African savanna, where our species evolved, saying "I'm no threat" clearly mattered more than "I don't get it." This is still true at many companies.

We have, of course, come a long way from macaques. Humans have evolved no fewer than fifty distinct smiles, some of them for highly specific functions such as flirtation. These biologically programmed expressions constitute a kind of universal human language. But our genetic heritage is never the whole story.

To complicate life further, different cultures have different rules for deploying this biological repertoire. When Japanese left fielder Hideki Matsui hit a grand slam home run in his first game as a New York Yankee, for instance, he rounded the bases with a sober face. In Japan, smiling would have been disrespectful to the pitcher. But he broke into a broad grin as he crossed home plate into the arms of his teammates, whose jubilant faces said, "Welcome to America. Dissing the pitcher's what it's all about."

Like Matsui, most people quickly master the right blend of biological signals and cultural rules for a given situation, and they come through the workday intact. But let's get back to Chip, who somehow got stuck with his fear grin, lips open, corners of the mouth drawn back in nervous anticipation. His knee-jerk reliance on this submissive display did not disarm or reassure anyone. On the contrary, it made them notice that he was frightened and insecure in his work. Chip was hired, it turned out, because his boss met him at a gym and thought his muscular build and Ken-doll looks "would be fun to have around the office." But he was utterly unqualified for his job. His grin-and-wave routine was an attempt to fend off anyone who might otherwise attack him for it.

Let's stay with the smile for a moment longer and further complicate the blend of biology and culture with the question of gender: Women are better at smiling than men. Though men are on average 15 percent larger, with proportionally larger muscles, women tend to have thicker zygomatic majors, the essential smile muscles running from the outsides of the eyes down to the corners of the mouth. No one knows if women are genetically prepared to be better at smiling. No one knows, that is, if the extra muscle was built in along with the fast-twitch fibers during the course of evolution. It is equally possible that this extra muscle is merely a cultural by-product of a lifetime spent smiling surly males into a less bellicose frame of mind. Possibly it's a combination of nature and nurture, like a born athlete given plenty of opportunity for exercise.

Regardless, smiling is something women tend to do more than men, often without realizing how the built-in overtones of submission or appeasement or even, to put it more positively, cooperation may affect their standing in the workplace. "It's an honest-to-God predicament," says Marianne LaFrance, a Yale University psychologist. "Smiling is clearly the default option for women. When they don't smile, people want to know what's the problem. When they do, they're in full gendered mode, and then people don't take them seriously."

There is no easy answer to such predicaments—except to be more aware of the raw material we are working with. Up to now, we have acted as if the biological component of our behavior scarcely existed. "Marx wrote of the importance of understanding the last few hundred years of our history," says Deborah Waldron, a professor of management and employee relations at the University of Auckland in New Zealand. "What of the thousands, indeed millions, of years of our forgotten history?" We need to be alert to the predispositions built into our genes during those forgotten eons, so that we can begin to steer them in a more prudent, humane, and even profitable direction. We can consciously steer these predispositions, or we can allow our inner monkey mind to blindly shape our behavior.

What I'm proposing in this book is a distinctly different perspective on working life, drawing on a variety of unconventional sources:

- It's based in part on the research of animal behaviorists and anthropologists, who typically look at their subjects in much the same way, watching for thousands of hours and recording exactly who does what to whom, and how often. Biologists have begun to blur the line between the zoo and the workplace, as evidenced by recent scientific papers with titles such as "Payment for Labour in Monkeys" and "The Chimpanzee's Service Economy." Economists have also contributed to this blurring, adapting animal foraging behavior, for instance, as a model for economic decision making.

Likewise, anthropologists are finding that field study techniques refined in Bongo-Bongo are eminently suited to deciphering the ethnography of Silicon Valley dot-coms. "Perhaps you have heard stories about a tribe called the Arioi in Tahiti," says corporate anthropologist Karen Stephenson. "Members of this tribe are divided into seven classes or grades, distinguished from one another by taboos which grow increasingly more complicated as one rises in the hierarchy. Whoever wanted to join this tribe would have to dress in an unusual style and behave as though mentally deranged. Does this remind you of any familiar business rituals?"

▪ Laboratory scientists add depth and complexity to the story by revealing what's going on inside study subjects while they are doing something or having something done to them. For example, economists now routinely collaborate with biologists using magnetic resonance imaging (MRI) to take pictures of the brain during game theory experiments. What sorts of things do they learn? One such study recently traced the appetite for a payday to the nucleus accumbens, an area of the brain also linked to drug and alcohol cravings. Other studies are sorting out how the brain responds to reward and why the anticipation of intermittent, unpredictable rewards may motivate people more powerfully than a steady paycheck.

Still other types of experiments have shown that a bad boss can cause subordinates to suffer elevated levels of cortisol, the so-called stress hormone. No surprise there, but what's more intriguing is that chronically elevated cortisol levels can cause cell death in an area of the brain called the hippocampus, with devastating effects on a person's well-being. If you are a beleaguered subordinate wondering whether to grin and bear it, or if you are an overbearing boss wondering why so many of your subordinates seem vaguely brain-dead, read on.

▪ Finally, the perspective I put forward in this book will use the work of evolutionists to help place the who-does-what in the larger framework of how individuals, companies, and species survive and prosper over the long term. For example, fear is a dangerous tool on the job. And yet it also seems to work, at least for a little while, and evolutionists can shed light on the when, where, and how.

I should acknowledge one other important element in this perspective: We liken ourselves, and especially our fellow workers, to wild beasts because it's fun, because it's colorful, and because laughing at ourselves (or at the boss, anyhow) is the best way to get people to swallow the unpalatable news that we really do act like big apes after all. With that in mind, this book will range as widely as possible through the animal world, from clownfish to redwing blackbirds, rather than limiting ourselves to our closest relatives among the apes. I take the view that humans fit into the broad spectrum of animal behavior (some of our coworkers more so than others), and that we can get a fresh look at our working lives by learning how even very different animals run their own particular rat races.

HAGFISH

What's the most accurate animal model for the brand of employee relations practiced by the new senior vice president for human resources (the SVP for HR, as we like to say)? One possible candidate: the bottom-dwelling hagfish, which likes to enter a victim by any available orifice, take up residence within, and eat out the victim's innards until there's nothing left but a hollow sack of skin. (Probably better not to say so out loud; she might like the idea.)

Weasels, you say? They are not nearly bizarre enough to explain the strange nature of our working lives.

2

NICE MONKEY
The Search for the Unselfish Gene

Managing is like holding a dove in your hand. Squeeze too hard and you kill it; not hard enough and it flies away.

—TOMMY LASORDA, BASEBALL MANAGER

Animal behavior research probably owes an apology to anybody who has ever been victimized by a self-aggrandizing business leader. (Yes, yes, I'm so sorry. Please take a number. The line today is only ten thousand miles long.) Ideas from the jungle and the savanna have had a profound influence on the behavior of great corporations. And given my premise that a natural-history point of view can be a helpful approach to workplace behavior, this ought to be a good thing, right?

Sadly, not always.

One problem is that scientists are human, too, and often find in the natural world exactly what their theories and predispositions incline them to find. Moreover, even the best biological ideas are often fragmentary and misinterpreted, not to say butchered, by the time they filter through to the boardroom and the corner office.

For a lot of businesspeople even now, the only thing from Darwin that matters is a phrase he did not invent, "survival of the fittest," typically invoked when compensation for top management takes the express elevator to the penthouse as low-ranking workers are being fork-lifted into the abyss. (Darwin borrowed the phrase "survival of the fittest" from the philosopher Herbert Spencer. But he never regarded sociopathic behavior as a mark of fitness, and he ridiculed those who did. "I have noted in a Manchester newspaper a rather good squib," he wrote to a colleague, "showing that I have proved 'might is right' and therefore Napoleon is right and every cheating tradesman is also right.")

The idea of executives as ruthless predators got a further boost from the scientific world in the mid-twentieth century. Biologists who had lived through the horrors of World War II commonly depicted early human ancestors as brutal, bloodthirsty killers. Every college student read *On Aggression,* in which Konrad Lorenz argued that humans are innately violent. Or if they didn't read it, they saw the "killer ape" idea embodied in *Lord of the Flies,* where schoolboy castaways degenerate into tribal savagery. They saw it distilled in the opening scene of *2001: A Space Odyssey,* in which some hairy ancestor slaughters his enemy and then tosses his bloody weapon, a bone, into the air, where it somersaults across the millennia and becomes magically transformed into a space station; all of human achievement was thus supposedly rooted in our violent nature. This underlying belief persists even now in the abundance of images of executives, management coaches, and even politicians putting on the full killer-ape pose, with arms folded and lips pressed together: *No nonsense. Me alpha. Mean business.*

RAGING EGOS

Over the past few decades, no biological idea has been more widely misinterpreted or had a more destructive influence on business behavior than "the selfish gene." In his 1976 book by that catchy title, Oxford evolutionist Richard Dawkins argued that we are little more than a product of our genes, and that these genes have survived by being as ruthlessly competitive as Chicago gangsters.

The title was a metaphor. Genes obviously are neither selfish nor unselfish; they have no motives. Dawkins merely meant that the basic business of a gene is to get as many copies of itself as possible into the next generation, by whatever means. He has protested ever since that he never meant to advocate selfish behavior as the best way to accomplish that. "Let us try to *teach* generosity and altruism," he wrote in his book, "because we are born selfish."

The idea of our innate selfishness, like the idea of survival of the fittest, had an intuitive logic, and some people latched onto it with glee. Since the time of Aristotle, critics had disdained merchants and other commercial types for being selfish bastards. Now, after two thousand years in denial, businesspeople suddenly said, "Oh. Right. And your point is what, exactly?"

"In the last twenty or so years, something very odd has happened," John Kay, a professor of management at Oxford, remarked in a 1998 speech. "This unattractive characterisation of business, previously put forward only by those who were hostile to it, has been enthusiastically adopted by businesspeople themselves. They no longer feel obliged to deny that their motives are selfish, their interests narrow, and their behaviour instrumental. They routinely assert that profit is the defining purpose of business activity." Kay didn't stick the blame for this on his Oxford colleague Dawkins. There was no shortage of other voices to blame, notably the University of Chicago economist Milton Friedman, who in 1970 declared, "The social responsibility of business is to increase its profits." But Friedman was merely saying what everyone expects a free-market economist to say.

Dawkins, on the other hand, was a biologist, and, his own intentions to the contrary, he provided businesspeople with reason to think that selfish behavior was more than a matter of economic convenience. It was also *natural*. And if it was natural, why be shy about it?

In fact, why not take it to every possible absurd extreme? Why not devote a company, Tyco, to the purpose of giving CEO Dennis Kozlowski's wife a birthday party in Sardinia, featuring an ice sculpture of Michelangelo's *David* pissing Stolichnaya? Why not make WorldCom a personal piggy bank for Bernie Ebbers? In this blinkered, egomaniacal spirit, one American broadcaster, a self-styled "scourge of the liberal media," recently argued that Michael Milken's selfish greed did more for humanity than Mother Teresa's selfless generosity.

But what if the idea of our innate selfishness is wrong?

Or maybe that's putting it a little too optimistically. What if we are born with raging little egos (as anyone who has ever tried to comfort a squalling infant at three in the morning will readily attest), but also just as innately born to satisfy our selfish needs by being social and cooperative, even altruistic? This isn't an easy idea to accept, particularly for businesspeople who feel that being "tough" or "hard-nosed" is the necessary price for success, or even survival.

To be frank, it isn't easy for me to accept either. When I last occupied a corner office, as managing editor of a magazine with an editorial staff of thirty, it was my job to be critical and demanding, and I often took it to the point of being abrasive, as old colleagues still frequently remind me with almost no provocation. My subsequent career as a writer has also inclined me to the normal journalistic fondness for harsh words, bad news, and good fights. When I write about business, it's typically about misbehavior. When I write about natural history, the subtext is usually violence.

As a writer for *National Geographic,* I have raced wild dogs to the bloody scene of a kill in Botswana, and once, in the Serengeti, witnessed cheetahs take down six gazelles in a single day. And I can assure you that this is the day I wrote about, rather than the previous three weeks spent watching cheetahs preen and loll about in the sun. "Nature, red in tooth and claw" may not be an accurate or representative view of animal life, but it's what people want to read. Like most authors, I also have a raging little ego and once published a book with the dedication "To hell with you all." So it's safe to say that my own predispositions did not incline me to look for cooperative behaviors, in the jungle or the workplace. But researching this book, and looking at humans and other animals with this book in mind, has made me think we have been overlooking the central fact of our working lives: Nature built us to be nice.

BLUE STEAM

The idea of being nice, or even strategically or manipulatively nice, may be hard to stomach in a business context. But it should not at first seem terribly surprising. Everybody knows the value of the personal touch, of making a connection with the people around us in the workplace. Smart companies have always incorporated the idea, however crudely, into their working lives.

Ken Wine, now a merchandiser at Lands' End, started out as a kid at a metal-plating company in Rhode Island: "The first thing they did was give you a five-gallon bucket labeled 'Blue Steam' and send you up to accounting to collect the blue steam. Accounting didn't have the blue steam, so they sent you over to manufacturing, and along the way, everybody asked who you were and what you did. It was a way of helping you get to know people and humiliating you at the same time. If you didn't have a sense of humor, you didn't belong there."

Likewise, when Ellen Moore started in marketing at W. L. Gore in Maryland, "they gave me a list of probably fifty people and said, 'You need to go around and get to know these people.' " She spent much of her first six months chatting with people, observing the work, "just hanging out." A time-and-efficiency expert would surely have scowled.

But one of Moore's assignments was to launch a new Gore-Tex product, called Pac-Lite, with the selling point that you could stuff it into a travel bag more easily. Two days before the launch, someone came up with idea of putting samples of competing fabrics into Pyrex cylinders, then adding a weight to each cylinder to demonstrate how much better Pac-Lite compressed. You are perhaps thinking that if Moore had spent less time schmoozing and more time being tough-minded about her work, this idea would have occurred long before her deadline. But what to do next?

"There was a maintenance guy who fixed machines around the plant. I used to chat with him. 'What did you do this weekend?' That sort of thing. I needed some weights made and I wanted them anodized so they would look nice going down in the cylinder. I told him the trade show was in two days. He got big eyes, because he had other jobs to do, and this needed some engineering. It had to be scientifically accurate. He said, 'For you, I'll make it work,' and he did, and I know it was because of that relationship."

In China, business and social networking depend on a carefully calibrated system of exchanging gifts and favors, called *guanxi* (literally "relationships"), which can influence everything from winning multimillion-dollar contracts to getting fresh fish in the market. And even in the smarmy, cutthroat world of American investment banking, some people understand that it pays to be nice, and not just to those above you in the hierarchy.

"The good associate recognizes from day one the value of good copy center relationships," write John Rolfe and Peter Troob in their book *Monkey Business,* about their time as Wall Street grunts at Donaldson, Lufkin & Jenrette. "The good associate greases the wheels of progress, even when there isn't an imminent need for express service in the copy center. The good associate orders up five or six pizza pies for dinner every couple of weeks and sends two of the pies up to the copy center. The good associate runs around the corner to the deli once a month, picks up a case of beer, and delivers it to the copy center guys. The good associate stuffs a twenty-dollar bill into the pockets of the key copy center guys at Christmastime and engenders some goodwill, or he stuffs a fifty into those same pockets and engenders twice as much goodwill. And then, when the need for express copy service actually arises, the good associate finds his job pushed to the head of the line while the badly mannered bitter associate spews forth futile vitriol, and gets the job returned three hours after his deadline. One hand washes the other."

SELFISH REASONS FOR BEING GENEROUS

Being nice—even strategically nice—is a natural behavior of monkeys, apes, investment bankers, and many other creatures formerly deemed savage. Monkeys typically live in groups of family, friends, and hangers-on, and they spend an inordinate percentage of their waking hours grooming one another. One individual will sit for twenty minutes at a time, stroking, caressing, picking gently through his partner's hair, and making a show of finding imaginary burrs and bits of dirt. The recipients of all this attention often stretch out and close their eyes in apparent bliss. They lounge around like pampered clients in a spa—a very nice spa. You start to envy them a little.

Biologists have spent much of the past forty years trying to devise a plausible explanation of why baboons, or businesspeople, would ever want to be nice, generous, or even altruistic with one another. They knew it happened, naturally. They just couldn't make sense of it, in the narrowest Darwinian terms. Doing favors, sharing food, and helping rear somebody else's offspring all entail giving away resources, whereas the stereotypical Darwinian hard-charger supposedly thinks only about accumulating them. Or as the bumper sticker on your idiot neighbor's Humvee puts it: "The one who dies with the most toys wins."

Biologists have generally attempted to explain altruism in selfish terms, which may seem odd. But being generous would never have become part of human nature in the first place unless it somehow benefited the individuals who display generosity. That is, a generous predisposition must have helped our ancestors survive and pass on more of their genes to future generations. Otherwise, altruism would gradually have disappeared from the human gene pool. On these terms, biologists have proposed three explanations for altruistic behavior:

- Reciprocal altruism holds that individuals do good things for one another in the expectation that the favor will eventually get returned. Grooming behavior, for instance, is how primates do their social networking, and it works: In a fight, a grooming partner is far more likely than a mere acquaintance to come racing to the victim's rescue. For humans in the workplace, reciprocal altruism generally does not start with picking ectoparasites off our colleagues. We do our grooming instead with friendly conversation and kind words, as Ellen Moore did at W. L. Gore, or with shared pizzas. Some biologists argue that humans evolved language in the first place as a substitute for grooming behavior, a new and improved form of social bonding. So it's probably not an accident that our most familiar statement of reciprocal altruism is "I'll scratch your back, you scratch mine."

- Kin selection holds that individuals do favors for their relatives, basically as a means of getting a piece of their own genetic heritage into future generations. A corollary is that our willingness to help is directly proportional to the degree of relatedness, or as a biologist once joked, "I would gladly die for two brothers, four cousins or eight second cousins." If you somehow imagine that this is not relevant to the workplace, bear in mind that 80 to 90 percent of all businesses are family-owned or -controlled. This includes about 185 members of the Fortune 500 (among them Johnson & Johnson, Marriott, Nordstrom, and Wal-Mart) and such Fortune Global 500 companies as BMW, SAP, Suntory, Hutchison Whampoa, and J Sainsbury.

Kin selection is what happens when a vervet monkey matriarch passes on her status to a granddaughter, or when the founder's great-grandson, William Clay Ford Jr., becomes CEO at Ford and thirty-three-year veteran Jacques Nasser gets shoved off the top of the tree (albeit with golden parachute deployed).

▪ Of the three explanations, the handicap principle is the newest and most counterintuitive. It reckons that animals and humans engage in risky behaviors, including altruism, basically to show that they can. The bigger and bolder the display, the more status it generates. The peacock's tail is the classic example of a handicap display. It's costly to maintain and so cumbersome it can make it difficult for the peacock to escape from predators. But when the male fans it out and rattles those absurd feathers, it's like Austin Powers saying, "Hey, baby, I'm still here."

It's easiest to see the handicap principle in the risky recreational pursuits of upper management—heli-skiing, race-car driving, piloting planes in mock dogfights. But heli-skiing and philanthropy are near kin. It was a handicap display when Massachusetts manufacturer Aaron Feuerstein opted to keep his employees on the payroll and rebuild after his textile factory burned to the ground, and it won him a reputation as the "mensch of Malden Mills." CNN founder Ted Turner was likewise making a handicap display and winning mensch points when he announced his intention to give $1 billion to the United Nations. (Unfortunately, Feuerstein soon went bankrupt, and Turner later saw his net worth plunge by billions of dollars, demonstrating just how risky an altruistic display can be.)

THE AFFILIATION INSTINCT

But there is another explanation for altruistic behavior, which applies at least as well to Feuerstein or Turner, and over the past few years, biologists, anthropologists, and economists have all been nervously dancing around it. Their idea is that we are not, after all, the rugged individualists we like to imagine. On the contrary, we tend to dwindle away on our own. (They mean this literally: People deteriorate more rapidly from heart disease, chronic stress, HIV, and other disorders if they lack social support. They may well have more toys, but they die younger.) It's only in the context of groups and relationships that we become fully human.

This is a discovery aging entrepreneurs typically make when they sell the company they have built up over a lifetime and suddenly realize that they are no longer alphas but isolates. Separation anxiety, a deep, disturbing sense of loss, also routinely afflicts retirees and workers who have been furloughed. It happens to cops when a partner of fourteen years suddenly gets transferred to narcotics.

Like most monkeys and apes, we are intensely social animals. We have an *instinct for affiliation.* When we affiliate ourselves, we tend, despite our stubborn intentions to the contrary, to mutate a little, to shapeshift, to mind-meld: Our emotions infect the people around us, and their emotions infect us, contagiously, instantaneously, unconsciously. We begin to dress like our new colleagues, talk the company talk, bleed blue for IBM. At Nike, some young sales reps call themselves Ekins (*Nike* spelled backward), and get themselves tattooed with the Nike swoosh. It's like the record store clerk in the movie *High Fidelity* who gets hired to work two days a week and then starts showing up every day. The workplace becomes our troop, our tribe, and we start to see the world in us-and-them terms.

For better or worse, the connection is visceral, even physiological: the people around us can influence our blood pressure, our production of biochemicals such as serotonin, dopamine, cortisol, and testosterone, our neural circuitry, even perhaps our ability to reproduce. In this context, what some biologists call the "tend-and-befriend" response is at least as natural as the more familiar fight-or-flight response. Being nice, nurturing other people, and being nurtured by them is as necessary to us as breathing clean air.

The evidence for this contrary idea has begun to surface from within our own brains.

QUIET JOY?

If work is a rat race, as we often complain, our dark secret is that we like our fellow rats. We tend to forget this, or maybe never realize it in the first place, because we get so caught up in the daily struggles, the frustrations, the unbelievable stupidities of the workplace. But people typically spend about 80 percent of their waking hours with other people, and we are built to enjoy keeping company. Forming social alliances and cooperating with coworkers gives us pleasure. It makes us healthier. It helps us live longer. And this isn't just because, as television-addicted toddlers, we warmed to the sight of Mr. Rogers putting on his cardigan sweater and asking, "Won't you be my neighbor?"

We know that cooperation is deeply rooted in our evolutionary history in part because the pleasure shows up in some of the most primitive emotion centers of the brain. In a famous experiment with bona fide four-legged rats, scientists planted electrodes in the anteroventral striatum, an area in the midbrain that's rich in receptors for dopamine, one of the brain's natural pain relievers. The electrodes enabled the rats to self-administer pleasurable sensations by pressing a bar. The rats liked it so much they would literally starve themselves to death rather than stop pressing the bar.

Stupid rats, right? But humans have a striatum, too, and friendly social contacts produce pleasurable sensations there. In a recent experiment at Emory University, behavioral scientists used magnetic resonance imaging, or MRI, to record activity in the brains of women engaged in social interactions. The women had the choice to cooperate unselfishly or to cheat. The context was a classic laboratory game called Prisoner's Dilemma: Two players meet briefly, then sit in separate rooms and play twenty rounds via computer. On each round, a player can choose a greedy option or a cooperative one. If both opt for greed, each gets $1. If one chooses cooperation and the other chooses greed, the first player gets nothing and the greedy one gets $3. If both choose to cooperate, they each get $2.

When players cooperated, the MRI images showed increased activity in the striatum. The women appeared to be experiencing the same neurological rewards from working collaboratively that the rats got when they pushed down on the bar. The test subjects themselves reported that they also felt better when they behaved generously and collaboratively. Science writer Natalie Angier nicely summed up the results: "Hard as it may be to believe in these days of infectious greed and sabers unsheathed, scientists have discovered that the small, brave act of cooperating with another person, of choosing trust over cynicism, generosity over selfishness, makes the brain light up with quiet joy."

Several small caveats: The scientists don't know for sure why it feels good. Dopamine may get released because something unexpectedly good has occurred, rather than because of the cooperative behavior. They also have yet to replicate their results in men, and it's possible that men's brains are wired differently, with a more obsessive focus on social dominance than on cooperation. Nor have they tried it yet in middle managers. So perhaps some ambitious executive trainee is already plotting how to use the Emory study to build a career at Ralph Lauren by lowering costs in Third World sweatshops: "The microchip on the sewing machine stitches up the shirt, right? So we connect it by wireless to an electrode in the operator's striatum and that way it can stimulate the dopamine receptors *at the same damn time.* It says right here they'd *starve themselves* rather than stop working." It's probably a good thing that the cost of brain surgery is prohibitive, for now.

But there's also behavioral evidence to draw a more hopeful conclusion: that male and female humans alike are built to cooperate. To be fair. To be moral, even. These tendencies may be at least as natural as selfish greed. "Morality is not a superficial thing that we added on very late in our evolution," says the primatologist Frans de Waal. "It relates to very old affectionate and affiliative tendencies that we have as a species, and that we share with all sorts of animals."

THE CHEMISTRY OF TRUST

Prairie voles, for example, are mousy little vegetarians native to the midwestern prairie states, and researchers regard them as a crucial means of illuminating the neurophysiology of the social bond. They help explain how we form social and sexual ties, and perhaps also what's going on when we experience the deep satisfaction of playing on a team or working in a group where things suddenly somehow click. They also offer a hint of what human resources is dimly getting at when it takes Busy People away from Important Work and subjects them to exercises in caring and sharing. ("Now just close your eyes and fall backward," your Team Spirit coach earnestly implores. "The members of your team will catch you. . . .") Even economists have begun to look at voles seriously, with the idea that their behavior may help explain what happens when negotiations go well. They may, for instance, explain what produced one of the most dramatic breakthroughs in the history of Middle East peace negotiations. But more about that in a moment.

Prairie voles are highly social animals, nestling together ten or twelve to a burrow. They are also deeply monogamous. Male and female form a permanent bond and, despite the occasional sexual indiscretion, stay together for life. Both also participate in rearing their young. By contrast, their close relatives, the montane voles, are cowboys. They don't have much use for company, except at mating time, and then—yee-haw!—they will have sex with anyone, anywhere. They also sometimes eat their young. ("Note to HR: We see managerial promise here.") Since the two species share 99 percent of their genome, Tom Insel, now head of the U.S. National Institute of Mental Health, wondered what makes their behaviors so different. He tracked it down to two peptide hormones, called oxytocin and vasopressin, produced in the brain.

Doctors frequently inject pregnant women with a pharmaceutical form of oxytocin, called pitocin, to induce intense uterine contractions resulting in childbirth. Oxytocin also enables the mother to let down her milk and nurse her young. But new fathers experience an increase in oxytocin, too, along with a surge in vasopressin. The physiological effects produced by these hormones often wind up becoming emotional and behavioral effects, too. Oxytocin lowers heart rate, respiration, and blood pressure, inducing calmness and a greater readiness for affection. Vasopressin seems to be connected with alertness and the urge to protect and care for the new family. All of these hormonal changes predispose new parents to form a social bond with their baby, who might otherwise seem like a loud, leaky little alien.

The same sort of hormonal preparation happens during sexual bonding (another case of loud, leaky aliens). Courtship starts the release of oxytocin, vasopressin, and dopamine in the male and female. Along with the physiological changes, each feels a fading of social anxiety and a growing comfort with letting the other come close. In prairie voles, these hormones are the key to forming the lifelong bond. When researchers inject more of the hormones into their brains, the voles fall for each other even harder. In montane voles, by contrast, an injection of oxytocin and vasopressin does nothing to alter their promiscuous behavior or otherwise transform their footloose cowboy personalities into something more settled and sociable. So the hormones alone aren't the answer.

It turns out, Insel found, that the two species differ dramatically in brain structure. In prairie voles, the neural receptors for oxytocin and vasopressin are more heavily concentrated in areas of the brain associated with reward and reinforcement. So when a male and female mate and experience the resulting rush of hormones, they become literally addicted to each other. The sight (or, more likely, smell) of the beloved apparently produces a blissful interaction between oxytocin and the feel-good neurotransmitter dopamine. In montane voles, the neural receptors are distributed elsewhere in the brain, and these voles get no kick out of you, or you, or you. One vole is as good as another, and social bonding is nil.

What about humans, a species also sometimes said to be monogamous? How much do our brains resemble those of prairie voles? Or are we montane voles in disguise? When researchers look at students who say they are madly in love, magnetic resonance imaging shows that their neural activity also appears to be concentrated in areas associated with addiction. In fact, their brains look remarkably like those of people using cocaine.

EMOTIONAL BARGAINING

But you probably are not madly in love with your coworkers. So how is all this hormonal research relevant to the workplace? Even between people who are strangers, small acts of trust produce a surprisingly rapid surge in blood levels of oxytocin. In an experiment at Claremont Graduate University in California, researchers randomly paired up anonymous individuals in a computer lab. The researchers gave "decision maker one," or DM1, $10 and told him that he could share any amount, from zero to $10, with "decision maker two," or DM2, who would actually receive triple that amount. Then DM2 could choose whether or not to give some of that money back, at which point the experiment ended. The two members of each team communicated only via computer, not face-to-face.

Standard economic theory—the model of "rational economic man"—predicts that, under these circumstances, the level of trust ought to be zero. The evolutionary theories of reciprocal altruism and kin selection also fail to provide any basis for expecting generous or trusting behavior. DM2 will never meet DM1 and can expect no benefit from being nice, nor any risk of retaliation for being a cheapskate. Realizing this, a DM1 should simply pocket the entire $10 and head for the door. But in hundreds of trials, according to Paul J. Zak, director of the Center for Neuroeconomics Studies at Claremont, 75 percent of the DM1s shared their money. Even more remarkably, 90 percent of the DM2s who received this sign of trust reciprocated.

Why were they being so nice? Despite the notion that we do business on a purely rational basis, neuroeconomists such as Zak suggest that our emotions often matter more than the facts. And the emotions involved are highly likely to be under the influence of oxytocin. In the Claremont study, blood tests revealed that the more the DM2s received from their DM1s, the more their oxytocin levels spiked, and the more they reciprocated. Zak concludes that oxytocin rises when someone trusts you, and *the act of trust by itself encourages trustworthiness.* Thus voles have raised the sneaking suspicion that human beings may well have an evolutionary propensity for trust. Fidelity, Mutual, and Beneficial may be more than wishful names on a company door. They may in fact be an expression of human nature, a product of living for eons in groups and tribes where trusting other members of the group was a matter of survival.

Once again, there are some caveats. On purely technical grounds, the question of whether blood levels of oxytocin correspond to brain levels is complicated, because of the "blood-brain barrier," a group of cells controlling which substances get to enter the brain. If you want to measure the actual oxytocin level in the brain, you need to drill holes in the skull and stick in needles, or make lumbar punctures, and most human test subjects are not quite that trusting. But animal studies tentatively indicate that blood and brain oxytocin levels do in fact correspond. Researchers have also found such a correspondence in direct measurements on the cerebrospinal fluid of autistic humans, whose inability to form social bonds may be influenced by an oxytocin deficiency.

A more substantial caveat is that laboratory experiments are not real life. The money in the Claremont study was pocket change, not, say, a $50 million construction contract. And even with the stakes so low, only 75 percent of the DM1s behaved generously.

What happened to the other 25 percent? Were the stingy ones simply rational economic types destined to become future middle managers? The DM2s also did not all behave in a trustworthy manner, their oxytocin levels to the contrary. One DM2, who became known around the lab as "Fat Bastard," was overweight and required four needle sticks before the technician could find a vein for his blood test. The researcher who interviewed him noticed that he was nonetheless "too happy." On being questioned, Fat Bastard boasted that he'd gotten $30 from his DM1 and kept every penny. His blood test showed a big oxytocin response, according to Zak. "But he suppressed it, he turned it off." Thus oxytocin may encourage trustworthiness, but it doesn't dictate it.

When the Claremont study made the news, the correspondents at one economist's Web site traded skeptical jokes about negotiating contracts only when the person on the other side of the table was on an oxytocin bender. A few of them speculated about future use of pharmaceutical oxytocin, or even "oxytocin-boosting beverages," as a way to induce fellow feeling in a group. (In fact, the blood-brain barrier makes this difficult with current methods.) One correspondent pointed out, a little cynically, that there is really no need to wait: "There's a good low-tech way to implement your plan. Next time you're in negotiations, shove an adorable wee baby into your opponent's face. You, on the other hand, concentrate on the stinking nappy."

He was kidding, of course. But in fact, the photographic equivalent once served as an effective device in bringing about the successful resolution of Middle East peace negotiations. On the thirteenth day of the Camp David negotiations between Egyptian president Anwar Sadat and Israeli prime minister Menachem Begin in 1978, Begin was refusing to sign the peace agreement. Bags were packed and everyone was preparing to go home disappointed. President Jimmy Carter, acting as mediator, took two steps that later proved critical.

First, he drafted a new version of a key document, removing a few nettlesome details without actually changing the substance. Then he sat down and signed commemorative photographs of himself with Sadat and Begin, which Begin had requested for his grandchildren. "Knowing the trouble we were in with the Israelis," Carter later wrote, "Susan [Clough, his secretary] suggested that she go and get the actual names of the grandchildren, so that I could personalize each picture." Then Carter went over to visit Begin on the porch of his cabin.

What followed was pure oxytocin-induced bonding: "I handed him the photographs. He took them and thanked me. Then he happened to look down and saw that his granddaughter's name was on the top one. He spoke it aloud, and then looked at each photograph individually, repeating the name of the grandchild I had written on it. His lips trembled, and tears welled up in his eyes. He told me a little about each child, and especially about the one who seemed to be his favorite. We were both emotional as we talked quietly for a few minutes about grandchildren and about war." Carter left behind his revised draft. Then he walked back to deliver the bad news to Sadat. As the two of them were talking, Begin called to say he had changed his mind and was now prepared to go ahead with the peace agreement.

It is of course simplistic to suggest that all it takes to bring people together in this fashion is a little oxytocin. We are only beginning to explore the neurophysiology of trust, and it will no doubt turn out to be far more subtle and complex than we imagine. The important point for now is that our biology gears us for cooperation at least as much as for conflict. Our default mode as social animals is not selfishness, but strategic altruism.

All this naturally raises a question: If social groups are so important to our identities, to our sense of comfort, security, and achievement, and if the workplace has become the main outlet for our social behaviors, then why do we go home so many days feeling burned out, dissatisfied, even downright miserable?

It was an ordinary morning on a dusty island in the vast, flooded Okavango Delta. A troop of baboons, and the naturalists who study them, had been walking vaguely all morning toward the shoreline, everyone alert for the ordinary African hazards of Cape buffalo, elephants, and lions. If the baboons decided to cross the water to the next island, we would have to wade in along with them. We were hoping it wasn't going to be one of those nasty days when a crocodile launches itself missilelike from the bottom. We were thinking it would be good not to startle a hippo and cause it to trample somebody underfoot.

The baboons gradually assembled on the shoreline, milling around and muttering "hunh, hunh," as if considering the same dire possibilities, and trying to decide whether to push forward or fall back. They had seen a member of their troop taken by a lion at this crossing a season or two before, and the alarm still sounded in a dim corner of their brains. Some of them climbed into the trees to gnaw on jackalberries and scan the water for danger.

It was what a baboon CEO might call an inflection point, a critical moment of decision making in the face of rapidly changing circumstances. Do we go forward into the unknown? Do we risk the possibility of predators? Or would it be wiser to withdraw?

In fact, what was happening among the baboons at the edge of the water that morning was not just a critical moment of decision making. It was the Primal Decision. A moment that had happened so often in our distant past, at so many forgotten water holes and river crossings, with so much trepidation and bloodshed, that it got burned into our brains. The way we once approached water holes now shapes the way we approach all decisions. It's what gives us a brain system (called negativity bias) that's always seeing ten good reasons not to go forward, that's constantly scanning for hints of a leopard lying in ambush.

On the shoreline, an elderly female with a reputation for fearlessly attacking leopards waded in up to her armpits. A wave of grunting passed through the troop, like the rumbling and naysaying of backbenchers in the House of Commons. Then an elder statesmen among the males stepped out in the lead female's wake, and others soon followed. Each hesitated. Each thought, Oh, man, I'm gonna get eaten alive out there.

Then, one by one, they stepped out into the still water.

3

BEING NEGATIVE
Why Things Look Worse than They Probably Are

Men are very queer animals—a mix of horse-nervousness, ass-stubbornness and camel-malice.

—T. H. Huxley

One day I was talking with my father about cooperative behaviors in the workplace, and he promptly launched into a string of anecdotes about New York City in the 1950s, including one about an outsize executive who, for reasons best not examined in great detail, liked to style himself Chief Copper Breast. "This was when we were still taking three-martini lunches," my father recalled, "and he loved to fling the metal drink trays and yell across any bar where we were having lunch, no matter how well the bartender had performed, 'Tell that cross-eyed motherfucker to fill this martini to the brim or I'll cut his balls off!' "

I laughed politely, then cleared my throat and said, "But I was asking about cooperative behaviors . . ."

"Oh," my father replied and fell silent for the first time in our conversation. "He always used to tip well," he ventured finally. "It's the only reason they let the dumb bastard come back the next day."

At first I attributed this response to individual psychopathology and too many three-martini lunches. But when I talked about cooperative behaviors with other people, people who are normal functioning employees of major companies, they also frequently fell silent. Or they began by talking about working as a team, then slipped unconsciously into the language of conflict.

At one Fortune 500 company, an executive described how she went to work as head of safety at a scientific research facility. She scheduled a get-acquainted visit with the security chief so that they'd be able to work together on her issues of fire safety, radiation safety, and the emergency response team.

"What do you want?" the security guy said when she showed up for their appointment.

"I just want to be part of the team," she began gamely.

"Let me tell you," he replied, "you're either part of the team or we roll right over you."

She grinned, or perhaps it was a grimace, and made cheerful conversation, just long enough to cover her hasty exit. Back in her office, she closed the door and tried not to imagine herself as a pancake. Colleagues advised her that the security chief was "a madman with a budget," a former state trooper who kept machine guns in his guard shack. She did not take much consolation in this news.

At home that evening, she began to devise a strategy for overcoming the negative vibrations. She felt a certain urgency about this mission: Among his other idiosyncrasies, the security chief's paranoid idea of burglary prevention was to keep the fire exits chained shut from the outside. But let's leave this unhappy impasse for the moment and come back to the safety officer's strategy later in the chapter.

THE PALENESS OF COMFORTS

We are built, it seems, to think in terms of conflict, not cooperation, and to a certain extent, it makes sense that this should be so. Cooperation may be the stuff of our daily lives, but precisely for this reason it can also seem boring. Ask a junior account executive what happened at work on a day when the company moved back into the black after three losing quarters, and she will probably say, "Not much." But when Suky steals a client from Kyle, this is something into which everybody can sink their conversational teeth.

Paul Rozin, a University of Pennsylvania researcher studying the problem of why people focus on the negative, refers to "the phenomenological paleness of comforts," meaning that "people generally don't get pleasure from their air-conditioning, but would experience immediate discomfort if it ceased to operate." Or as Schopenhauer once remarked, "We feel pain, but not painlessness."

It is, in fact, our biological nature to accentuate the negative, to notice the one dumb thing that goes wrong rather than five or ten things that go right. We differentiate between negative and positive events in just a tenth of a second, and the negative ones grab our attention.

For instance, when researchers show test subjects a paper with a grid of smiley faces on it and one angry face, the test subjects instantly zero in on the angry face. Reverse the pattern, and it takes them much longer to pick out the solitary smile. Likewise, when a boss makes four positive comments and one quibble in an employee review, the subordinate almost invariably fixates on the quibble. Neurologists call this tendency "negativity bias," and they regard it as a survival mechanism. A heightened focus on what can go wrong helps us deal with danger. An angry face grabs our attention more urgently than a smile because it represents a potential threat.

Negativity bias got built into our minds during millions of years of evolution because animals and early humans who didn't have it got a brief, bloody lesson in natural selection. The little Tommy gazelle that went around blithely grazing with his head down sooner or later looked up to find himself in the middle of a pride of lions: "Ah, good morning, old chaps. What big teeth you—"

Excessive blitheness tended to get cut short and thus became less and less common in succeeding generations because, as Rozin delicately phrases it, "the threat of a predator is a terminal threat." Staying alive meant keeping your eyes and ears open, constantly looking up to scan for ghosts in the tall grass—the stealthy leopard, the trio of hungry cheetahs, the skulking investigator from the Securities and Exchange Commission. Skittishness, or negativity bias, became a distinguishing characteristic of the survivors.

In fact, we have evolved at least three separate neural systems concerned with escape from danger. At the level of the spinal cord, we draw back reflexively from pain. At the level of the limbic system, we deploy an animal repertoire of escape and defense reactions. And at the level of the cortex, we dwell thoughtfully on the hairy, scary possibilities that seem continually to beset us.

On the other hand, if we spent all our time being skittish, we'd never get out of bed. We'd never start a new business. We'd never go to work. Or if we did, we'd shut the door and hide under our desks to avoid all the problems, a behavior not unknown among new managers. So neurologists theorize that evolution has also equipped our brains with the opposite tendency, a "positivity offset," encouraging us to approach rather than to withdraw, and thus enabling us to ask somebody out on a date, or apply for a big job, or elbow our way to the bar.

The approach and withdraw systems are apparently as separate as the parts of the brain that handle vision and hearing, according to University of Chicago researcher John C. Cacioppo. Moreover, they are built to function at the same time. "Coactivation" of negativity bias and positivity offset is an evolutionary by-product of the daily need to go about our business in the face of continual danger. "In the savanna, for example, animals must come to the water to drink even though predators come there to hunt," writes Cacioppo.

An individual equipped to think about the appetizing possibilities and the scary ones at the same time, rather than one after the other, is much more adept at making nuanced decisions about when to tiptoe closer, and when to run like hell. The really successful gazelle is the one with sufficient negativity bias to react instantly when a lion poses a genuine threat, and enough positivity offset to approach and drink when it discerns that the lion is just another animal with a full belly idling around the water hole. Thus it prospers in a landscape of fear.

WHY NEGATIVITY MATTERS

The idea of negativity bias first appeared in academic circles in the 1960s, though Rozin points out that the disproportionate power of the negative had previously turned up as a fact of life in sources from Shakespeare ("The evil that men do lives after them; the good is oft interred with their bones") to Russian folklore ("A spoonful of tar can spoil a barrel of honey, but a spoonful of honey does nothing for a barrel of tar"). Negativity bias has lately attracted renewed interest as neurologists and experimental economists both attempt to get at the roots of apparently irrational human behaviors. The evidence increasingly suggests that this evolutionary predisposition continues to have an enormous influence on the way we run our lives and our businesses in the modern world.

For instance, it is the bane of pollsters and marketing executives with their focus groups that people routinely say one thing and do the opposite. In a 1993 Gallup poll, for instance, 85 percent of Americans approved the idea of becoming organ donors. But only 28 percent had actually signed up to become donors. Why the discrepancy? Cacioppo blames it on negativity bias—in this case, the fear of dying combined with the irrational suspicion that donors in the emergency room risk being harvested prematurely for spare parts. At some primal level, doctors in this scenario become the threatening equivalent of hyenas at the watering hole.

Once we are aware of negativity bias, Cacioppo suggests, we can counter it by educating people better, emphasizing that medical staff don't do that sort of thing, and rarely even know whether a living patient is an organ donor or not. Or we can make donor status the default mode, so people have to opt out rather than opting in. The United States uses the opt-in method and—negativity bias in action—lets sixty-five hundred people die each year on transplant waiting lists. Worldwide the annual death toll is twenty thousand. A dozen European countries, including Spain and France, do it the other way and suffer far fewer deaths due to the lack of a donor organ.

Negativity bias also helps explain why we suffer exaggerated fear of economic loss but experience relatively little emotion about profit. We remember Black Monday from 1987 and even Black Thursday from 1929. But, quick, where is White Wednesday or Bright Friday? Vanished, apparently, on a cloud of exultant stockbrokers swathed in phenomenological paleness.

The exaggerated focus on loss helps explain why an economic downturn loses votes for the incumbent party during U.S. presidential elections but an economic upturn has no noticeable effect. It also sheds light on why employees may bitterly resent being asked to give up a job benefit even though they didn't much want it in the first place. Researchers have shown that if they randomly give people a possession and then, a few moments later, ask them to sell it back, they value it at around twice the price they would pay if they did not already possess it. Experimental economists suggest that the disproportionate aversion to loss is a basic psychological tendency, which routinely causes people to make irrational economic decisions.

Presented with the same basic economic proposition, for instance, we make markedly different choices depending entirely on whether the questioner frames the proposition in terms of gain or loss. Let's say a researcher walks up and offers you either $1,000 or a 50 percent chance of getting $2,500. If you're like most people, you will choose the $1,000 sure thing. But if the researcher offers you either a $1,000 loss or a 50 percent chance of losing $2,500, people mostly choose the riskier option. That is, we seem to be conservative when it comes to increasing our gains. But we are so alarmed by loss that we will take considerable risk to avoid it.

Understanding this built-in bias can make a life-or-death difference to a business. For example, the willingness of one employee to take extreme risk in a vain attempt to erase a loss brought on the downfall of Barings Bank. In 1995, the bank's star Singapore derivatives trader, Nick Leeson, lost U.S. $1.2 billion trying to recoup a $317 million trading loss he had incurred. Barings subsequently collapsed, after 250 years in business.

Experiments suggest, oddly, that negativity bias *doesn't* affect our perception of intelligence, which may partly explain the tendency of superiors to misjudge someone like Leeson. It may also help explain why organizations that pride themselves on their intelligence sometimes blunder overconfidently into the dumbest possible mistakes. "A person who behaves highly intelligently on one occasion, and stupidly on three occasions, is still seen as intelligent," according to Rozin. He explains it in terms of the scarcity of intelligence: Somebody may have one good idea a year and five bad ones, but good ideas are rare and can make you money, so you stick with him. "In terms of ability, you judge people by their best performance."

In terms of morality, on the other hand, negativity bias kicks right back in and we judge people by their worst performance. For a social species like ours, a sign that someone is untrustworthy is almost as significant a threat as a predator. So the taint of wrongdoing looms disproportionately large and lingers long afterward. Understanding this tendency in the public mind is critical in the aftermath of a corporate scandal or a major product flaw. Even if the scandal was the work of only one or two executives, and even if the product flaw affected only a tiny fraction of what was sold, it can still do enormous damage to a company's reputation. "People treat these as very powerful events, like murders," says Rozin.

The way a company deals with this instinctive public tendency to fixate on the negative can produce dramatically different results for its reputation, even given the exact same offense. In 2000, for instance, software maker Oracle got caught hiring an investigative firm to paw through the garbage of public advocacy groups supported by archrival Microsoft. Oracle whined that it had ordered the firm not to do "anything illegal," and CEO Larry Ellison declared it his "civic duty" to investigate what the company called "Microsoft front groups." The deny-and-defend strategy earned Oracle a *Wall Street Journal* editorial for its "slimy antics."

Less than a year later, Procter & Gamble chairman John Pepper discovered that his company had been doing much the same thing. Moreover, unlike Oracle's spies, the P&G spies actually got the goods when they raided the garbage at the hair care division of Anglo-Dutch rival Unilever. Pepper set out to deliver the unmistakable message that at P&G this was not business as usual. He not only fired three executives involved, he also revealed the episode to Unilever. P&G eventually agreed to pay Unilever $10 million in compensation and to refrain from using any of the information it had illicitly acquired. This was at a moment, moreover, when P&G was deeply unpopular on the stock market and had dropped tens of billions of dollars in net worth.

Despite the natural tendency of consumers to remember bad behavior, P&G thus held on to its straight-arrow image as one of the best companies in America. It also got honored in a *Fortune* "Best and Worst of 2001" article for its candor. But while the P&G incident was quickly forgotten, the words "corporate espionage" still conjure up the vision of Larry Ellison Dumpster-diving in his Armani shades.

Negativity bias can also do damage by encouraging individuals or companies to become obsessed with the competition only as rivals. Ellison, for instance, has declared, "We pick our enemies very carefully. It helps us focus. We can't explain what we do unless we compare it to someone else who does it differently. We don't know if we're gaining or losing unless we constantly compare ourselves to the competition." But his fixation on Microsoft was personal; the two companies had relatively little competitive overlap at the time of the garbage incident. In the real world, moreover, enemies are also sometimes allies, and a company, like a gazelle at the watering hole, will do better if it can make shrewd judgments about when to approach and when to retreat.

In their influential 1996 book *Co-opetition,* business professors Adam Brandenburger and Barry Nalebuff point out that people often focus so passionately on cutthroat competition that they become blind to areas of mutual benefit (or even to whose throat is being cut). Citibank, for instance, pioneered the automatic teller machine in 1978. "When other banks came along with their own ATMs, they wanted Citibank to join their networks," the two authors write. "That would have made everyone's ATM cards more valuable. When banks are on a common network, each machine complements all the others. But Citibank refused to join. It didn't want to do anything that might help its competitors. It didn't want to help Dr. Jekyll if that also meant helping Mr. Hyde." The other bank networks soon came to lead the market, and Citibank was finally obliged to join them in 1991, after considerable inconvenience to its own customers.

Business isn't, after all, about beating the enemy. It's about making money, and that means figuring out the best course in a shifting landscape of cooperation and competition. Brandenburger and Nalebuff attribute the obsession with competitors as enemies to the "business-as-war mindset." But the roots of negativity run much deeper.

IS YOUR ENEMY ALSO YOUR ALLY?

In Israel's Negev Desert, the density of edible seeds in the soil can vary a dozenfold from one patch to the next. So a smart gerbil ought to forage in the patch with the most seeds, right? But if the patch is out in the open, the gerbil could end up as dinner for a barn owl. If the patch is under the shelter of a bush, on the other hand, a sand viper could be lying in wait. The nervous gerbil must choose where to get its livelihood based on incomplete information about an incredibly complex and treacherous marketplace, where the penalty for getting it wrong is instant death. Sound familiar? University of Illinois researcher Joel Brown calls it "an ecology of fear."

But look at it from the predators' point of view: Owl and snake, though apparently competitors, actually function as allies: "The snake frightens gerbils into the talons of owls, and owls frighten gerbils into the fangs of snakes." Like Wal-Mart and Target playing cat-and-mouse with a hungry manufacturer.

Rivals also prove useful in what biologists term the "dear enemy" effect. In many species, from songbirds to foxes, territory holders avoid conflict with neighbors. They seem to understand that fighting over the property line with the guy next door is a distraction intruders will exploit. Instead, these "dear enemies" may actually band together to perform "coordinated evictions" when some upstart tries to horn in on the neighborhood. Archer Daniels Midland, the so-called supermarket to the world, practiced an illegal form of the "dear enemy" strategy in the mid-1990s, conspiring with rivals on five continents to jack up the price on essential ingredients in a host of supermarket products. One of the company's unofficial mottoes was, "The competitors are our friends, and the customers are our enemies."

WHY CAN'T WE ALL JUST GET ALONG?

When people show up at the watering hole nowadays, even in the corridors of the very best companies, they often pass their time grousing. It hardly matters how sweet the life may look to an outsider. At *National Geographic,* writers and photographers traveling on assignment in some of the most spectacular landscapes on earth often pass their free hours trading stories about the idiocies being perpetrated back at headquarters on M Street in Washington, D.C. Or they quarrel with each other. On M Street, meanwhile, staffers mutter about the idiocies being perpetrated by egomaniacal writers and photographers in the field. And they quarrel with one another, too.

Organizations often aspire to be like big happy families, but they succeed mainly in the sense that they spend much of their time bickering. This is negativity bias at its most nitty-gritty, day-in-day-out level, where it colors every dealing with every individual other than ourselves. (Oh, why not include ourselves, too? We all have those dark moments when we stand alone in the bathroom, staring at the mirror and thinking, "I'm not with that person." And even animals sometimes chase after their own tails.) Stress compounds our fractious tendencies. We antagonize people even when we do not really mean to, because we always seem to have too many places to get to and too much to do in too little time. Or as a Pepsi executive once eloquently phrased it, "We've all been so busy, we're like a Chihuahua trying to bury a bone on a marble floor."

What to do? How to fret and bicker less? How to live a little more positively? There is good news and bad news. The bad news for hard-nosed managerial types is that the remedy for squabbling and bickering sounds at first a lot like a tune from a Broadway musical, specifically "Happy Talk" from Rodgers and Hammerstein's *South Pacific*. The good news is that it is at least *quantifiable* happy talk. The people who study such things agree about precisely what the ratio of positive to negative interactions ought to be for a successful workplace. At or around the magic ratio, groups overcome negativity bias and work together productively. Below it, they bicker, brood, and fail.

Marcial Losada, an organizational psychologist, spent ten years at H. Ross Perot's Electronic Data Systems (EDS) closely observing different management teams as they developed their annual strategic plans. EDS put up $20 million buying high-tech gear and building special meeting rooms with one-way mirrors to help Losada make his observations. Losada was doing exactly what any primatologist would do: watching large, dangerous animals from the sidelines and reducing their complex interactions to a vast numerical database of behaviors.

Observers reviewed videotape of the meetings and counted positive behaviors (like saying "That's a good idea") and negative behaviors (like saying "That's the plumb dumbest thing I ever heard"). Other researchers independently categorized teams as low-, medium-, or high-performance, using actual results for profitability, customer satisfaction, and peer approval. Then Losada's group put these two sets of results together.

It turned out that the fifteen high-performance teams averaged 5.6 positive interactions for every negative one. The nineteen low-performance teams racked up a positive/negative ratio of just .363. That is, they had about three negative interactions for every positive one.

The obvious conclusion is that, because people fixate on quibbles, slights, slurs, contemptuous gestures, skeptical facial expressions, and perceived threats, it takes a lot of stroking to reassure them and build trust. At one family company, the vice president and heir apparent is an attractive woman with big sea-blue eyes, blond hair pulled back, and a pearl necklace, and she is acutely, almost maternally, sensitive to the way a sharp word can cause subordinates to draw their heads in like turtles.

"If I see body language that indicates that, I will work very hard to take the turtle back out of the shell. They'll physically shut down." She whispers the sound effect: "Waaa-*chuk!* Often the head will lower and the shoulders will slump and they'll lean back in the chair and let you know they're out of this game. You need to immediately recognize that you have done that and say, 'Oh, look, you have some really good points, we've got to develop them.' And provide a safe environment." Now she makes her voice comically small. "So the little turtle head will come . . . back . . . out. *Very* gently."

Listening to this in a coldly rational business setting is a little hard to take, and yet also, on some emotional level, a little hard to resist. Maybe it's those big blue eyes.

"Being a family member," she adds, "everybody watches you. That's okay, that's what people do. So it's a wonderful opportunity to create an incredibly positive environment. *Choo!* That's the best. That's the most fun."

It isn't enough, though, merely to accentuate the positive. In fact, Losada found that when the positive interactions waxed much beyond the five-to-one ratio, teams again began to grow ineffective. Negative interactions were necessary to keep things healthy. They served as an essential reality check, though "Here's why I think that might not work" was clearly better than "Oh, you incredible flea-brain."

So it is certainly a mistake to regard a team as a failure simply because the members of the team argue. Conflict and disagreement frequently lead to progress. Or as Orson Welles put it in *The Third Man:* "In Italy for thirty years under the Borgias, they had warfare, terror, murder, and bloodshed. But they produced Michelangelo, Leonardo da Vinci, and the Renaissance. In Switzerland, they had brotherly love; they had five hundred years of democracy and peace. And what did that produce? The cuckoo clock!" Other researchers have found that conflict is good in the early stages of decision making, bad in execution; good when a task is new and intellectually engaging, bad when the task becomes routine.

In all circumstances, context is paramount. What makes a group of humans work together successfully, Losada emphasizes, is the *ratio* of positive to negative, of good vibrations to static.

Getting the ratio right is essential not just because of negativity bias but because we are such social and emotional animals. Despite what we tell ourselves, it is never just about getting the job done or picking up the paycheck. For instance, other researchers have demonstrated that feedback alone can increase productivity by 10 percent. This may seem merely logical, since feedback clarifies both the task and the expectations. But getting *acknowledged* for accomplishments can enhance performance by 17 percent, apparently for the entirely emotional reason that it builds trust and reinforces the bond between the individual and the group.

Social tools such as acknowledgment and approval are beneficial "even for high-paying organizations that believe they are inundating their managers and employees with monetary rewards," according to University of Nebraska business professor Fred Luthans. Monetary incentives alone may increase performance by 23 percent. But when managers combine the money with feedback and social recognition, they can get a 45 percent jump in productivity.

The idea of the five-to-one ratio may still seem a little daunting for the average workplace. This is an awful lot of positivity. What's even scarier is that Losada's five-to-one ratio also appears to be essential when you get home and try to muster the energy for a successful marriage.

John Gottman at the University of Washington has found that couples with a ratio of fewer than five positive interactions for every negative one are destined for divorce. Curiously, the magic number also seems to have a close parallel in the ratio of positive behaviors (grooming, sharing food, sitting together) and negative behaviors (biting your rival's scrotum) among monkeys and apes. Thus the five-to-one ratio begins to look suspiciously like a basic primate need.

BETTER CAGES FOR CORPORATE ANIMALS

Since the 1960s, zoos everywhere have been working to figure out how to make caged animals feel at home. That typically means getting them out of their narrow cages and into a habitat with more sunlight, plants, and other positive elements from their native ecosystem.

Now what about us? The average cubicle monkey gets cooped up under fluorescent lights, with windows that do not open, or no windows at all, in a suburban office park, where the word *park* apparently refers to parked cars. It's enough to make our inner beast curl up in a corner and suck its thumb. Corporate animals, like zoo animals, have a built-in genetic need for some connection to the natural world. Hope may lie in something called the green building movement.

Green buildings often look much like any other building, at least superficially. When a visitor arrives at the headquarters of the Genzyme Corporation in Cambridge, Massachusetts, for instance, no signs announce that it is a green building. You could make a sales pitch there, do an interview, sign a contract, suck down coffee from a foam cup, hustle off to your next appointment, and never be any the wiser. It's just a handsome glass box with a twelve-story atrium, lots of handsome plants, and office space for nine hundred people.

Stand still long enough and you may notice that the high-tech venetian blinds are adjusting themselves a few degrees at a time to shield you from glare. When you enter a restroom, the lights flick on and a sheet of hand towel shoots out in readiness. Heliostats on the roof track the sun and radiate natural light into almost every corner of the building, via a system of mirrors and matte-surface reflective ceiling tiles. The artificial lights overhead automatically dim themselves as the sunlight waxes, then brighten again as it wanes.

But you don't need to know about all that, or even care.

And this nonevent is really the big news about the green building phenomenon. Largely forgotten or treated with open derision just a few years ago, the idea of a building in tune with the environment, where energy conservation and the comfort of the resident animals are top priorities, has suddenly become mainstream. In Philadelphia, Comcast is building a green skyscraper 975 feet tall. In mid-Manhattan, Bank of America is building one with a filtration system that removes 95 percent of particulate dust from air entering the building. In Washington, the state government has declared that it will build nothing but green. Even Wal-Mart and Target are building what purport to be green "big box" stores.

So what does it mean to be green? Is it easier to spend an average twelve-hour workday there? Is it even possible for a building 975 feet tall to be green? How about a suburban office park? Is the term *green building* itself an oxymoron?

This new wave of green buildings does not conform to a particular green aesthetic, probably a good thing given the ugly duckling sensibilities of the first generation of green buildings in the 1970s and early 1980s. Modern green buildings tend not to employ bunker-style berms, in-your-face solar arrays, or other ostentatiously environmental techniques. They also eschew the first-generation proclivity for righteous discomfort, on the theory that a building cannot truly be green if people don't like being there.

Modern green buildings simply do their job more efficiently and intelligently, and they now earn their green bragging rights by the numbers. The U.S. Green Building Council, a partnership of builders and environmentalists, sets standards through its Leadership in Energy and Environmental Design, or LEED, program. Among other things, LEED awards points for better ventilation, greater use of natural lighting and views, better thermal comfort and control, and the use of paints, sealants, carpets, and other materials that minimize toxic emissions. The total points add up to qualify a building for a LEED-certified, silver, gold, or platinum rating. Over the past four years, the Green Building Council has accorded green status to 188 buildings, with another 1,800 in the pipeline. By one estimate, the United States will have 10,000 LEED-registered buildings within the decade.

What's caused this resurgence in a type of construction once thought to be as dead as the Clivus Multrum composting toilet?

"These buildings have absolutely stunning performances—with average savings of 30 to 70 percent for energy use, and 50 percent for water use," says Rick Fedrizzi, founding chairman of the U.S. Green Building Council. "They deliver great bottom-line value for owners. They treat the people who use the building with respect. And, oh, by the way, it's good for the environment." The pragmatic language of costs and benefits has proved more persuasive than environmental rhetoric. "Build green," the council now urges. "Everybody profits."

But no single factor has given a bigger boost to the movement than the idea that green buildings help people live and work more productively. Studies suggest that patients in green hospitals recuperate faster, shoppers in green retail spaces spend more money, students in green classrooms test better, workers in green factories suffer fewer injuries, and overall productivity rises by anywhere from 6 to 16 percent, depending on the study. The idea that people do better in green buildings makes sense, says Fedrizzi, because they are no longer cooped up all day in sealed, climate-controlled boxes breathing toxic fumes. One study calculates that the traditional way of building, without regard for environmental considerations, costs the nation "on the order of $60 billion" a year in productivity lost just to so-called sick building syndrome.

Some advocates of the green building movement say most such studies so far are too anecdotal. "There's nothing rigorous you can take to an industry and say, 'Here's why you want to do this,' " says one engineer. What's really needed, adds an architect specializing in green buildings, is a study that asks: "If you live and work in green buildings, do you use your health insurance less?" (To judge the existing studies for yourself, go to the research section of www.usgbc.org.) Still, the common impression is that green buildings simply feel better.

"This is the first building I've ever been in that was built for people," says Rick Matilla, sitting out on a planted terrace in the atrium at Genzyme, where he is director of environmental affairs. "It doesn't feel like you're inside a building. The air feels fresher, there's natural light, there's a sense of transparency and connection to nature. I'm surprised we didn't grasp that concept a long time ago—getting out of caves and going outside."

NEGATIVITY ON THE BRAIN

In a laboratory at the University of Wisconsin in Madison, a volunteer lies on his back with his head in the donut-hole opening of a magnetic resonance imaging machine, which pings and squeals as the huge, hidden magnet coil spins round, taking a picture of his brain in thirty slices. It is a little unnerving to see the living patient on one side of the glass wall, and the image of his disembodied brain on the other, rotating in 360 degrees, or tilting up and back at the operator's command.

The MRI is recording the life of the patient's mind. To be more precise, it is recording blood flow and other metabolic data to measure the activity in different areas of his brain at any given moment, areas of increased activity theoretically requiring more blood for the delivery of oxygen. Among many other things, MRIs allow neuroscientists to locate negativity bias in the brain.

When people are emotionally distressed—anxious, angry, depressed—most of the brain activity takes place in the right prefrontal cortex, and in the amygdala, an area of the midbrain that is roughly Fear Central. By contrast, when people are in a positive mood—upbeat, enthusiastic, and energized—those sites are quiet, and the heightened activity occurs predominantly in the left prefrontal cortex.

University of Wisconsin neuroscientist Richard Davidson connects activity in the left side of the prefrontal cortex to "a whole package of behaviors" including the way we point, move toward an object, handle it, and then give it a name. The left side came to specialize in these approach behaviors during our evolution on the savanna, when they were crucial in deciding to advance toward a water hole or enter new territory. The right side of the prefrontal cortex, on the other hand, specialized in withdrawal behaviors, particularly detecting threat and backing away from it.

Every individual has an "emotional set point," according to Davidson, an individual tendency to approach or withdraw, and the MRI is a way to index it. At one extreme, people with a significantly higher level of activity on the right side of the prefrontal cortex are most likely to have a clinical depression or anxiety disorder over the course of their lives. At the other extreme, those with a distinctly higher level of activity on the left side rarely experience troubling moods and tend to recover from them quickly. Most people fall in between, with a familiar mix of good and bad moods.

The old lore held that the adult brain was essentially unchangeable, except for a remorseless decline as old neurons dwindled away with age. But Davidson has found that a person's emotional set point need not be permanent. Just because a boss overreacts to problems doesn't mean she has to spend the rest of her life snapping at subordinates. Just because a computer programmer shrinks from conflict, or even from contact, doesn't mean he is limited to enjoying a relationship only with his Samsung flat-screen. Scientists have recently discovered that even elderly people can produce fresh new cells in the hippocampus, an area of the brain critical for learning and memory. Davidson now describes the brain as "more designed to respond to experience than any other organ in the body." The list of factors within our control that can produce physical changes in the brain includes exercise, cognitive therapy, drugs such as Prozac, and Davidson's favorite mind-altering technique—meditation.

A MANTRA FOR THE MONKEY MIND?

Meditation may seem at first particularly unsuited to a world where everybody else wants to be an eight-hundred-pound gorilla. Or as I put it to Davidson, a little negatively, "Does any company really want employees to model themselves on Buddhist monks who accomplish nothing?"

"Do you consider the Dalai Lama someone who accomplishes nothing?" Davidson snapped. He had a photo of himself on his office wall shaking hands with the Dalai Lama, both of them bowing, Davidson slightly lower. So I refrained from suggesting that His Holiness might be too evolved to fit in at Honda, say, where workers have been known to chant, "We will crush, squash, slaughter Yamaha!"

Davidson has, however, tested the effects of meditation in one highly competitive workplace. Volunteers at a large Wisconsin chemical reagents company called Promega trained for half a day a week using traditional meditation techniques: sitting quietly, hands resting on laps, breathing deeply and becoming calm. They also practiced meditation at home for forty-five minutes a day.

"This culture is obsessed with certain practices, going to the gym to achieve demonstrable effects on the body," says Davidson. "But there is every evidence that if we care for the mind in the same way we care for the body, positive emotions like generosity, happiness, compassion, can be trained up. They are skills, not fixed characteristics."

Eight weeks into the Promega experiment, MRI tests showed that test subjects who had been practicing meditation experienced a 10–15 percent shift in the ratio of brain activity, away from the right side, bastion of negativity and withdrawal, and over to the positive, forward-thinking left side. The regimen also appeared to improve test subjects' physical health. Individuals who showed the greatest shift in brain activity also displayed the biggest improvement in immune-system function, as measured by antibody production. Davidson did not attempt to measure whether meditation improved productivity in more conventional workplace terms. But the experience of business people who use meditation is at least suggestive.

One of Davidson's test subjects, a senior scientist at Promega named Michael Slater, reports that the program actually increased the stress level at first, because busy people suddenly needed to find forty-five minutes a day to meditate. The process of paying more attention to what was going on in their own heads also ultimately caused some people to leave the company. "A few realized they were out of sync, maybe living out somebody else's agenda," Slater says.

His own experience, on the other hand, was that meditation produced a healthy change. He describes himself as a Type A personality, a worrier, with a streak of the Type T thrill seeker; he enjoys windsurfing and riding a motorcycle too fast. With meditation, he says, he found it easier to suspend judgment and practice "active listening" when meeting with people. "I don't react as much if my buttons are getting pushed. Instead, I ask why this is bugging me, and then I choose what to do about it. It maybe takes a half a second. It's not a big internal dialogue." He describes it as "not letting the monkey mind carry you away to a state of agitation, that constant chatter that's in most people's heads."

Slater equates greater calmness with increased confidence, and he states it in terms of the primal choice between approach and withdrawal: "I think it's always a benefit to not shy away from conflict, to not withdraw, to see failure as an opportunity for further expansion. I don't see how anyone could not perceive the approach mode as better for a business whose intent is to grow."

Well, okay, maybe so. But when Davidson espouses generosity, happiness, and compassion, a skeptic naturally wonders what job description those words fit these days, except maybe as traits that are not required. Meditation just seems a little too dreamy.

So it is worth considering one other workplace case study. When Bill George joined Medtronics in 1989, the Minneapolis company was a pacemaker manufacturer with a market capitalization of $1 billion. By the time he retired as CEO in 2001, the company had a far broader product line and was widely regarded, in the words of *The Economist,* as "the most innovative and market-savvy firm" in the intensely competitive medical device industry. Its market value had increased to $63 billion, a rate of growth to rival that of IBM in its prime. *Business Week* named George to its list of the top twenty-five business managers, and the Wharton School ranked him among the top business leaders of the past quarter century.

George, who now serves on corporate boards and in various nonprofits, clearly has plenty to keep him busy. But he has always found time to meditate, usually twice a day. He does it after breakfast, sitting with back straight and hands in lap, sometimes at home, sometimes on a plane in the twenty minutes before takeoff. He does it again in late afternoon and says this refresher gives him the energy to pick up work again later that night and stay sharp until midnight. He has a mantra, though he falls back on it as a prop only in times of high tension.

George started meditating in 1975, when he was a young company president with a long agenda of things he wanted to accomplish and a tendency to push subordinates too aggressively. He was, by his own description, "impatient, intimidating, and tactless," traits he now hastens to describe as having both positive and negative aspects for an organization. Meditation merely enabled him to "own" this forceful approach "and modulate it and moderate it." It gave him a sense of clarity and focus on the important things, so he could accomplish them without undue stress.

At Medtronics, he made it the goal to get 70 percent of company sales from products introduced within the previous two years, and in a single year, after a history of no major deals, he also oversaw the acquisition of six other companies at a cost of $9 billion.

Does this sound like enough stress? To stay relaxed, George says, he also jogs four or five days a week and goes in for regular massages. But meditation is more restful. Whereas massage takes out tense spots in the body, meditation takes out tense spots in the brain. "The media tend to make meditation some weird deal," he says. "I don't know why. To me it's a very natural thing. If this were a drug, it would be considered malpractice not to distribute it."

APPROACHING THE POSITIVE

Among the many dumb ideas about the natural world enshrined in business lore, there is one that goes like this: "Every morning in Africa, a gazelle awakens. It knows it must run faster than the lion or it will be killed. Every morning in Africa, a lion awakens. It knows it must run faster than the gazelle or it will starve. It does not matter whether you are a lion or a gazelle. When the sun comes up, you'd better be running."

The author of this inspirational saying was anonymous, probably out of embarrassment. He had almost certainly never been to Africa, where predator and prey live in close proximity and engage in a continual dialogue for the apparent purpose of avoiding needless running.

Sometimes a herd stands around alternately grazing and being vigilant, their bodies all pointed like compass needles toward the predator resting off in the bush. If the predator stands up, they all move back precisely beyond the predator's attack distance, to obviate a potential ambush, and then they graze some more.

Sometimes a group of animals will cry out to the predator, as if to say, "We see you, big boy. Don't think you can play Pop Goes the Weasel with us." One or two animals may even walk right up and inspect the enemy, as if to determine its intentions. When a chase begins, some antelope will leap high into the air right in front of the predator, a show-off behavior known as stotting. The message to the predator is, "Oh, *puhleeze.* Don't even bother. I'm too fast for you." Predators generally get the message and chase the animal that hasn't got the stuff to stot.

The bottom line is that anyone in business who spends the day perpetually running is a victim of fear and negativity bias. And sooner or later, he will get eaten alive by somebody who knows better. (He may even do everyone a favor and run himself right into the jaws of his enemy.) The way to survive isn't to live in fear but to survey the landscape calmly, alert but dispassionate. The way to achieve that state of calm awareness may involve exercise or meditation; it may be simply a matter of getting out of the office for lunch, leaving work at a reasonable hour, developing a life outside the job. (Physiologist Robert Sapolsky has found that baboons reduce the physiological signs of stress by playing with the kids.) Then, if there are hazards when you get back to the workplace, you face up to them on your own terms.

MADMEN WITH BUDGETS, DISARMED

In that spirit, our head of safety sat in her office at her new workplace thinking about how much she wanted to be part of the team and how fervently she wished that the team would not roll right over her. As in any workplace, the temptation was to yield to negativity bias and ignore the problem, in the hope it would go away. She didn't want to provoke needless antagonism with the head of security. And yet the buildings under her care were full of laboratories where researchers did volatile experiments. And she couldn't help thinking that, in a fire, the chains on the emergency exits would prevent them from getting out. It wasn't as if the chains were even necessary. The buildings were all contained within their own campus, protected by chain-link fencing, guard posts, and possibly also machine guns.

That night, the head of safety and her husband role-played the conversation she needed to have with the head of security. "I thought it was going to be a total pissing match," she admits. But she also knew that "achieving anything is far easier if you understand the other person's perspective. That's how you get cooperation even out of people who thrive on conflict." So together they came up with an opening strategy.

Next day, she sat down with the head of security again and said, "You're looking at building security from the outside in, and I'm looking at it from the inside out."

The transformation this remark elicited was so complete and gratifying that, years later, it still fills her with wonder. "It completely changed the tenor of the meeting. I manipulated him by getting him to see my perspective, because I knew his perspective. He said, 'There's this new technology. It's an emergency exit that locks from the outside, but with a paddle connected to an alarm on the inside. We can do that.' "

Thus the head of safety became part of the team, which did not roll over her. The problem was not only faced, but corrected. The chains slipped off the emergency exit, a sweet victory for the spirit of cooperation.

And everyone lived to bicker and whine for another day.

Negativity bias is one reason so many species practice herd behavior with such blind, fretful enthusiasm. During a visit in the Antarctic, for instance, Peter Brueggeman of the Scripps Institution of Oceanography sat in the middle of an Adelie penguin traffic jam watching the birds dither at water's edge. "So what does it take for them to jump in?" Brueggeman wondered in his online journal. "They watch the water and when a large group of penguins comes swimming into their immediate area, the Adelie penguins start getting very vocal." They jostle, jockey for position, squabble, peck back and forth, bash one another with their flippers, and engage in raucous discussion, followed by "an immediate chain reaction of everyone rushing to jump in the pool all at the same time, no waiting."

Why all this commotion? The penguins have grounds for negativity: Hungry leopard seals and orca whales patrol these shorelines in search of penguins for dinner. But the penguins need to eat, too, and sooner or later, if they can muster the positivity offset, that means going into the water. If there are already lots of penguins in the water, chances are that the coast is clear. So the dithering crowd on shore all try to do the same thing, jumping in at the exact same moment. Some of them may still get eaten. But this is the reward for social conformity: Chances are it'll happen to the other guy.

4

ROUGH BEASTS
Moore's Law Meets Monkey Law

Andy was big bark, little bite, and I was little bark, big bite.

—FORMER INTEL CEO CRAIG BARRETT, OF HIS PREDECESSOR, ANDY GROVE

When Andy Grove ruled Intel, he made a reputation for being intelligent, articulate, and disciplined. Casey Powell, a former Intel executive who was the victim of one of Grove's worst managerial blunders, still describes his former boss as "unbelievably effective, absolutely brilliant . . . This is a guy who, one-on-one with anybody, can reach into your chest, take out your heart, and talk to it." But on a bad day, Grove could also take out your heart and leave it in slices on the floor.

Grove himself boasted that he built Intel on a diet of fear and paranoia, "fear of being wrong and fear of losing," fear as "the opposite of complacency." He argued that he meant to focus this combativeness on outside competitors, and he later warned against making middle managers so fearful they were afraid to deliver bad news. But the Intel culture of "constructive confrontation" could also inflict wounds within the company. In his zeal for beating up the competition, Grove sometimes found himself ripping apart his own underlings.

In the early 1980s, he launched a program called Operation Crush, intended, as one staffer put it, to "snuff Motorola." When Motorola inexplicably survived, Grove came back a year later with Project Checkmate. Casey Powell, who was already putting in eighty- and one-hundred-hour weeks as general manager of microprocessor operations, also got the job of running Checkmate. But Powell soon realized that Checkmate wasn't moving fast enough to satisfy Grove. Crush had been a companywide fight for survival, with applications, marketing, and engineering people formed into "SWAT teams" and deployed around the world to win sales. But even with Motorola making a comeback against Intel's latest product, the 286 computer chip, it was hard to muster the same level of fear and intensity.

Powell recalls that he also wound up reporting to an intermediary, Jack Carsten, who despised him and seemed to be persuading Grove to feel the same way. Powell's feelings toward Carsten remain bitter, even decades later: "He absolutely beat up on the people who worked for him. He humiliated them publicly. If you kissed his ass, he would take care of you. But he would still do it, just to remind you." Carsten, on the other hand, says he and Powell are "still friends" and dismisses published accounts of their dispute as "a total fabrication."

Powell says Carsten soon asked for an executive staff review of Checkmate. A week beforehand, another senior executive warned Powell: "Grove's going to kick your ass." When Powell walked into the boardroom, it was evident that the other executives arrayed around a U-shaped table also knew. Powell, standing in the middle, started to describe some of the problems Checkmate was encountering.

In his unauthorized company history, *Inside Intel,* journalist Tim Jackson recounts that Powell's voice trembled, and Carsten immediately pounced: "Well, aren't *you* the problem?"

"Carsten—*ka-choong!*—shoots the first bullet in from the side," Powell recalls. "That lit Grove off, and he laid into me. 'I put you in this job, and I'll take you out of there . . .'

"It was so bad. . . . I graduated from the U.S. Merchant Marine Academy, and it felt like I was back in plebe year. I just stood there, stone faced, at attention, and listened to it. I was absolutely furious."

Grove quickly worked himself into a tirade, roaring accusations of incompetence and worse. Even by the standards of Intel's combative culture (where another executive used to show up for meetings wielding a baseball bat "to direct the discussion in a productive fashion"), the outburst was startling for its vindictive fury. A senior vice president eventually stood up and said, "This is bullshit. Andy, if you want to do this to the guy, take him someplace and do it. Don't do it here in front of everybody else." When another executive threw down his pen and started to walk out, Grove said, "Okay, I'll stop." He turned to Powell again: "Tell me what you're going to do to fix the problem."

Powell pulled himself together and went through the rest of his presentation, including the solutions he was putting into effect.

"Okay, fine," said Grove.

"Does that satisfy you?" Powell demanded.

"We'll just have to wait and see," said Grove.

"Okay, now let's talk about *my* problem," said Powell, and he could feel the tension in the room go *fwoosh,* up to the ceiling. "It's pretty clear that I am a 'does not meet' as a manager." It was a standard phrase for anyone falling short of company standards. "Right, Jack?" he asked, and he recalls that Carsten immediately said, "Yup."

Then Powell added, "It's also pretty clear I'm a 'does not meet' as a husband and as a father. Now. Here's the problem." He had spent his entire life exceeding standard expectations, and something had to be terribly wrong with Intel for it to seem otherwise. "So how are you going to solve my problem?"

"Well, what do you think I should do?" said Grove.

"Oh, you want *me* to solve the problem? I can solve the problem just like that." He snapped his fingers. He was glowering at Grove now, as if there were no one else in the room. Everyone recognized the sense of danger.

Someone said, "Stop this meeting now," and Grove quickly called a recess. In the aftermath, a few executives gathered around Grove and admonished him for his handling of the situation. Others expressed their embarrassment to Powell.

As Powell came around the table to leave, Grove was standing in the doorway and the two came face-to-face.

"I am sorry," said Grove. "I really am sorry I did that."

Powell looked at him. "I didn't know whether to cry or hit him. I said, 'I knew you were going to do this a week ago. How can you tell me you're sorry?' " Then he walked out of the building, collected one of his daughters, and went off to spend the rest of the day at a local amusement park trying not to think about work.

Six months later, Project Checkmate delivered exactly the results Powell had promised. Grove sent a gracious note acknowledging his work.

Then Powell left Intel forever, taking along seventeen employees to start his own computer company, Sequent, which soon grew into a billion-dollar business. He bought his chips, in the beginning, from National Semiconductor, not Intel.

"Andy thought I did it to get even," Powell says now. "I did it to *show him.*" He thinks about that for a moment, then adds, "Still, he's the guy I admire more than anybody else I ever worked for."

CHIMPANZEE POLITICS

Intel was arguably the greatest high-technology company of the personal computer era. And yet every move that day, every nuance of confrontation, followed the same rules that apply among rival male chimpanzees in the forest: Any chimp would have recognized the alpha's routine aggression as a tool for intimidating subordinates, the beta male's political maneuvering to weaken a rival, the subordinate's show of weakness as an invitation to attack, the glowering, the tirade, the use of an aggressive display in lieu of violence, and even the failed attempt at reconciliation by Grove, followed by the subordinate's inevitable dispersal to form his own troop.

Intel had written the signature rule of the computer age: Moore's law was the self-fulfilling prophecy that computer chips would double in memory every eighteen months. But its executives were playing by rules of primate behavior that date back thirty million years.

In fact, we all play by those rules.

In the same year as the Checkmate upheaval at Intel, an obscure Dutch researcher named Frans de Waal was casting a whole new light on the human workplace by opening up the social world of chimpanzees. It's a measure of de Waal's influence beyond the scientific community that his laboratory at Emory University in Atlanta has since been visited, though with unknown effect, by one of the richest alpha males in the world, Microsoft chairman Bill Gates. *Business Week* has featured de Waal's work under the headline "Manager See, Manager Do," with a sidebar on "management secrets of the chimps." And the *New Yorker* and the *New York Times* both prominently cited de Waal's work on fairness among primates to explain the fall of the scandalously overpaid New York Stock Exchange chairman Richard Grasso in 2003.

There is an arch, tongue-in-cheek quality to much of this reporting. But de Waal's work also resonates at a much deeper level. Harvard University biologist E. O. Wilson credits de Waal with "moving the great apes closer to the human level than could have been imagined as recently as two decades ago." In the process, de Waal has inadvertently also demonstrated how surprisingly little our own behavior has advanced beyond that of our fellow apes.

De Waal got his start as a young biologist at Arnhem Zoo in the Netherlands in the 1970s, when researchers still carefully avoided attributing feelings, thoughts, or even individuality to mere animals. The ban on such anthropomorphism was intended to discourage naively projecting human states of mind onto other species. But to de Waal, the idea of an absolute divide between the behaviors of animals and humans was paralyzing, a kind of "anthropodenial." It turned the animals into robots, "blind actors in a play" that only we understood.

As he sat watching and meticulously recording chimpanzee behaviors over thousands of hours, it seemed to de Waal that animals didn't act like that in real life. One of his favorite chimps, for example, was a deposed alpha named Yeroen whose blustering shows of dominance no longer impressed because he needed to sit down afterward "with eyes shut, panting heavily." Even in his dotage, Yeroen was enough of a schemer to play the younger males off one another and hang on as a kingmaker. He allied himself with a younger male named Nikkie, helping him to become the alpha. In repayment, Nikkie indulged the old fox's sexual forays with females in the group, a privilege the alpha would normally try to reserve for himself.

Yeroen probably should have been grateful. But to keep the boss on edge, he would also sometimes side with Nikkie's rival Luit. He was anything but a blind actor in this drama. (It hardly needs saying that the Yeroen strategy is common in the business world, too, sexual conniving included. Boeing, for instance, brought Harry C. Stonecipher out of retirement to regain a sense of probity after financial and sexual scandals under the previous CEO. "In pulling Harry out of the mothballs," newspaper columnist Colin McEnroe later commented, "the board must have been thinking: 'He's old. He bears a strong resemblance to Honus Wagner's glove at Cooperstown. He's not going to be tearing his pants off and jumping the help.' " Unfortunately, Stonecipher also could not control his Yeroen impulses and quit on account of a sexual peccadillo early in 2005. But let's stay with hairier apes for a moment longer.)

In the beginning, like most twentieth-century biologists, de Waal viewed chimpanzee life as an endless round of fighting for power and privilege. But he quickly saw that victory went not just to the strong. Watching animals such as Yeroen helped him discern that chimpanzees schemed for power by grooming one another, doing favors, cultivating useful alliances with family and friends, and other more or less wholesome social behaviors. When they fought, they often reconciled soon after, to keep the peace. What made an animal an alpha was not size and strength alone, but social skill. "Whole passages of Machiavelli seem directly applicable to chimpanzees," de Waal wrote in *Chimpanzee Politics,* an account of his years at Arnhem.

That title was originally suggested by *Naked Ape* author Desmond Morris, who was a friend and mentor. Morris recalls that the word *politics* was unorthodox in animal studies then, and de Waal resisted it at first but "finally had to agree that that was indeed what he had been studying." If Machiavelli's *Prince* had been the first book to frankly describe the power motives and manipulations in human hierarchies, *Chimpanzee Politics* was the first to show that these behaviors were deeply rooted in our animal evolution.

Among the "innumerable and interminable" social maneuvers observed by de Waal:

- High-ranking chimpanzees sometimes schemed together (like Nikkie with Yeroen) to discredit a potential rival. They also routinely recruited other apes into coalitions (like an office worker taking aside a potential ally after a meeting). They cultivated these coalitions to seize power and to maintain it against challengers. Subordinates also used coalitions to moderate the behavior of those in power. If an alpha male proved too brutal, coalitions of females could band together and force him out by siding with a rebellious rival.

▪ Chimpanzees sometimes engaged in exaggerated displays to manipulate the behavior of others. After Nikkie injured Yeroen in a fight, for instance, Yeroen made a point of limping—but only when Nikkie was in the vicinity, as a sort of pacifying or perhaps guilt-inducing reminder. Another time, Nikkie threatened his rival Luit from behind, and de Waal saw a nervous fear grin cross Luit's face. Luit put his hand up to his mouth, pressed his lips together, and only then turned around to face his rival—like an executive putting on his game face before entering a difficult meeting.

▪ Chimpanzees occasionally engaged in violent struggles within the troop. But (much as in the Intel meeting about Checkmate) other individuals often intervened to prevent a fight from spinning dangerously out of control. At Intel in the early 1980s, the top executives were almost all men. In this regard the chimpanzees were somewhat less primitive: females often figured prominently as peacemakers and as members of powerful coalitions.

In short, many of the behaviors we had previously classified as "office politics" were in fact primate politics. The idea of chimpanzee politics attracted widespread interest, not least among reporters, who asked questions like "Who do you consider to be the biggest chimpanzee in our present government?"

De Waal declined to make such comparisons. "People do it to mock the politicians," he remarks now, "but I feel they insult my chimps."

On the other hand, politicians themselves sometimes saw the resemblance. When he became Speaker of the U.S. House of Representatives in 1995, Newt Gingrich placed *Chimpanzee Politics* on his list of recommended reading for incoming Republicans. Gingrich himself proved adept at the alpha-ape business of fierce infighting. But he seems not to have paid attention to the parts of de Waal's book about using reconciliation and other social behaviors to hold the group together through all the conflict and maneuvering.

On the contrary, his political action committee produced a notorious memo advising Republicans to describe Democrats, regardless of circumstances, as "sick," "pathetic," "bizarre," "twisted," and "traitor[ous]." The lesson he apparently thought the chimpanzees were teaching was that it was possible to dispense with civility, when in fact what de Waal had discovered was just how crucial civil behavior can be, even among brutish chimps.

Gingrich was eventually overthrown as a result of precisely the sort of coalition building and rebellion de Waal had described among chimpanzees. (Later, Gingrich became an important offstage adviser—Yeroen to Secretary of Defense Donald Rumsfeld's Nikkie—in the George W. Bush administration.)

De Waal and Harvard biologist Richard Wrangham are now the leading voices for two starkly contrasting views of primate behavior that have emerged in recent years from chimpanzee studies. In his 1996 book *Demonic Males: Apes and the Origins of Human Violence,* Wrangham and coauthor Dale Peterson make the case that chimpanzees and humans alike are male-dominated species, territorial, with a built-in penchant for staging bloody and often lethal party-gang raids on neighboring communities. (This is the book Gingrich apparently thought he was reading.)

But de Waal went on after *Chimpanzee Politics* to emphasize peacemaking behavior in primates, their innate tendency, after fighting with other members of the group, to reconcile with each other. Wrangham's social order is more rigidly authoritarian, built on systematic brutality toward underlings. De Waal's social order, while also sometimes brutal, is more egalitarian. Female coalitions play an influential role, and there's a focus on cooperation and the shared values that hold the community together.

Curiously, these two viewpoints closely match two basic approaches sometimes espoused by management theorists for running large corporations. "Theory X" companies operate on fear and conflict. "Theory Y" companies tend to be more cooperative, more inclined to involve all workers in shaping the workplace. Both viewpoints have at least two things in common: Whether they are nasty aggressives or nice to the point of being nauseating, all companies and all chimpanzee groups experience conflict. And however egalitarian they may pretend to be, all of them also have powerful hierarchies. The two things are closely connected, though not in the simple cause-and-effect way we generally assume.

AN UNNATURAL FEAR OF CONFLICT

Apart from the military, most human organizations treat aggressive behavior as a social taboo. Human resources staffers typically want criticism to be positive. They want everyone to be nice. They want us all to *just get along*. When trouble crops up, they hire consultants to "facilitate" meetings and note down comments on one of three flip charts. The first flip chart is for comments about "what worked well," or WWW. The second is for "even better if," or EBI. The third flip chart is the "parking lot," a limbo for reluctantly acknowledging the unruly comments of malcontents who insist on pointing out that the whole thing was a giant freaking fiasco that cost the company $30 million.

"O-*k-a-a-y*," the consultant replies gamely, after a brief, agonized pause. "Now, how can we put that constructively?" Anything to get us back into the nonjudgmental territory of EBI.

The trouble is that, by themselves, being nice and just getting along seldom produce results. Or as London Business School professor Nigel Nicholson put it in a recent issue of *Harvard Business Review,* "In fact, one sign of a failed encounter—yet another 'Sure, boss' meeting—is the employee managing to get out of the room without expressing a contrary point of view." Or the subordinate gets to express a contrary point of view, but not about the issues that really matter. The boss lets employees go back and forth about minor details, to give the appearance of having a thorough airing of opinions—but without risking actual conflict.

One frustrated employee describes a boss whose "painful but very effective" strategy is to spend "the most time answering the questions from the dumbest people at the meeting so the people who might ask the hard questions just want to get out of there and get back to work."

The idea that people in healthy relationships religiously avoid conflict and aggression does not square with real-world experience. Most of us at some point have witnessed or participated in marriages, friendships, sports teams, and work groups where the effort to be unnaturally nice was debilitating and others where a combative atmosphere was wholesome. Consider two alternative examples:

▪ On one government commission, the chairwoman is aggressively polite, particularly toward another commissioner whose intelligence is not always obvious because of her baby face and soft voice. At a public hearing recently, Babyface announced that she would be handing off some of her duties to one of the other commissioners. The chairwoman leaned forward and, speaking slowly, said, "Are you *sure* this is the right thing to do? I *worry* about you, honey. Is this *really* what *you* want to do?" Babyface, being accustomed to chronic condescension, was about to let this roll by. Then she noticed the horrified jaw-drop expression on another commissioner's face, and suddenly she recognized her own suppressed feelings of outrage. She was still too polite to do anything about it. But the jaw-drop commissioner sent an angry e-mail to the chairwoman, who then phoned Babyface to apologize in tears. "She used tears because she thought that's what would work on me," says Babyface. "And I would've bought the tears if they had led to genuine change in behavior." But at the next meeting, the chairwoman was condescending all over again.

▪ At another workplace, the top boss regularly shouts at meetings and sometimes says things no manager should normally say. And yet the spirit of these meetings is buoyant, collegial, and robust, according to a middle manager who works for this benevolent despot. There is give as well as take, on all sides. The middle manager admits that it sometimes irritates her being made to look bad in front of her own subordinates. "So I'll turn to him, because I know he spent a year in France, and I'll say, '*Je t'encule.*' " Loosely translated, this means "Stick it up your ass," though in a somewhat more personal sense. "And we'll both laugh, and that breaks it up."

A human resources facilitator would probably have trouble finding a way to put this constructively. We are well beyond EBI here, and probably even out of the parking lot. But constructive is exactly the effect this remark has on the cortisol levels of everyone at the table. And the team seems to work together far more happily in its unruly way than the hyperpolite government commission ever will. The difference, of course, is that they like each other, they have a genuine relationship, and when they fight, they reconcile. It may seem odd, but humans, like chimpanzees, are bumptious creatures for whom an honest obscenity can sometimes work as a form of bonding where tears and insincere emotions utterly fail.

Managers often avoid conflict because of their own weakness or insecurity. They fear that if they encourage a reasonable level of debate, things will slip too easily across a dangerous line. Blunt talk, like all other forms of aggression, can cause hard feelings, as it did that day when Andy Grove worked himself into a rage against Casey Powell. We may try to focus the talk on goals and issues rather than on personalities. But we are emotional animals, and such discussions easily become personal. Losers feel their heightened cortisol levels simmering underneath their eyeballs. And the shouting and turmoil cause a frisson of unrest to race down the hallways and surge through the entire office.

But ducking conflict, or trying to avoid a difficult decision, is just negativity bias in action. The job is to approach conflict in a straightforward way, with a minimum of emotional chaff. Intel thrived in part because all disagreements got a full and often forceful airing. The company never attempted to gloss over internal dissent, though it did ultimately expect a unified effort behind the chosen course of action. The standard phrase in the company culture was, "I disagree, but I commit."

This is more or less what chimpanzees in a troop say to each other every day of their boisterous lives.

CONFLICT IS NORMAL

One day not long ago, de Waal sat watching his study animals from a boxy yellow tower beside an open-air compound, part of the Yerkes National Primate Research Center at Emory University in Atlanta, where he is a psychology professor. The compound was an area of dirt and grass half again as large as a basketball court, enclosed by steel walls and fencing. The chimps lounged around on plastic drums, sections of culvert pipe, and old tires. Dividing walls angled across the open space, giving the chimps a chance to get away from one another, much as do the cubicles in an office. (The walls, said de Waal, "let subordinates copulate without getting caught by the alpha.")

It was a warm, lazy southern afternoon. Below, one chimp strolled past another and dealt out a slap that would have sent a football tackle to the emergency room. A second chimp casually sat on a subordinate. Others hurled debris, charged, bluffed, and displaced one another. One chimp let out a resounding, outraged "Waaa!" and others joined in till the screaming swirled up into a cacophony, then died away again into torpor and mutual grooming—like a business meeting late in a catastrophic quarter.

De Waal, going gray at the temples, in round, wire-rimmed glasses and a "Save the Congo" T-shirt, smiled down on the apparent chaos in the chimpanzee compound. "Growing up in a family of six boys, I never looked at aggression and conflict as particularly disturbing," he said. "That's maybe a difference I have with people who are always depicting aggression as nasty and negative and bad. I just shrug my shoulders and say, 'Well, it's a little fight. As long as they don't kill each other . . .' "

De Waal notes that aggressive or antagonistic behaviors take up only about 5 percent of a chimpanzee's day. But he also doesn't attempt to minimize the importance of aggressive behaviors for chimps maneuvering to get ahead in the troop. On the contrary, he embraces their aggressiveness. Where other biologists and sociologists have routinely described aggression as antisocial and destructive, an expression of individual psychopathology, de Waal calls it "a well-integrated part of social life . . . It occurs in the best relationships."

In fact, de Waal describes the fight over mother's milk at weaning time as "the first negotiation in a young mammal's social life." The infant, finding himself denied the comfort of the breast, may pout, whimper, or scream, all aggressive behaviors. The mother may push him away from her breast or rebuke repeated demands angrily. As in every other relationship for the rest of our lives, a conflict of interest has inevitably arisen, and aggression is merely one tool for attempting to resolve it.

But mother and child have far more common interests than conflicts. So they also reconcile. A chimp mother may allow the infant to suck on her lip or her earlobe, much as a human mother gives the weaning infant a pacifier. They coo and tickle and hug, and this "cycling through positive and negative interactions" results in agreement about the terms of the relationship. De Waal calls this "the relational model" or the "conflict resolution model" of social behavior, and the relevance to the human workplace is obvious: We argue and find a new way forward. We disagree but commit.

De Waal's work acknowledges that we have evolved to use aggression as a normal "tool of competition and negotiation." More important, it also suggests how we can handle it prudently and still work together tomorrow. The important thing, he has found, is not whether the chimps fight, but "the perceived value of the relationship and the way the conflict is dealt with." That is, how they treat each other before and after.

And among chimpanzees and corporate types alike, that gets complicated depending on who's in charge.

BREAKFAST OF SIBLINGS

When the squabbling among your coworkers becomes insufferable, console yourself with the thought that you are at least doing better than the sand tiger shark, which practices intrauterine sibling cannibalism: The first little shark to hatch in Mama's uterus goes around gobbling up the other eggs and hatchlings—its brothers and sisters. This gives the shark the nutritional benefit of eating its fellows for lunch, and also thins out the competition for maternal resources. If your fellow workers aren't behaving at least a little more cooperatively than this, take the hint: Find a new womb, get a new job.

5

DONUT DOMINANCE
Why Hierarchy Works

Dogs look up to you. Cats look down on you. Give me a pig. He just looks you in the eye and sees his equal.

—Winston Churchill

In a small town on the coast of New England, a shack named the Beach Donut Shop opens its front shutters every summer to supply fried dough to the masses and employment to local teens. A seasonal donut shop offers its workers few opportunities to do combat for mastery of the universe. Instead, they fight over how to pump raspberry jam into a jelly donut, and the best way to stack sprinkled crullers on a tray. It is a prelude to the working lives that lie ahead.

"Jeff and Dylan take any excuse to assert their dominance over each other," a coworker reports. "The other day I walked into the back room to find the two of them staring intently at a tray of cinnamon buns. I knew it was going to get ugly. Our cinnamon buns are large, swirly pastries, which we smother with vanilla frosting. But on this tray, half the buns had a thin layer of frosting spread over the top, and half had a thick glob of frosting in the middle. Jeff stood next to the thinly spread buns, Dylan next to the thick ones, like corners in a boxing match."

"What's the idea?" asked Jeff, pointing at Dylan's side.

"That would be how you frost a cinnamon bun, my friend," said Dylan.

"I've never seen cinnamon buns frosted like that. And I've worked here longer than you."

"Obviously you didn't learn much in that extra time," said Dylan. "Those things look like total—"

"The customers don't care what they *look like.* They just want a ton of frosting."

"Don't be stupid. Of course they care. Weren't you just bragging about the aesthetic appeal of your raspberry donuts and how much the customers would love them?"

There was a moment of awkward silence. Then Dylan stepped away, gave a melodramatic bow, and said, "Touché, my friend, touché. I have to go fill some orders."

The witness to this scene comments: "Jeff's snarl subsided into a grim smile. In his eyes, his cinnamon buns had proved his donut dominance. But I knew this battle would rage on. As for myself, I was content to watch their petty conflicts, knowing my superiority in the donut world was undisputed."

Boys, boys, boys! Life in the donut shop should not be like that (especially since, for all the male infighting, one of the girls was really in charge). But, alas, we know it is like that, and we have a pretty good notion that ten or twenty years hence, these same three will be fighting alarmingly similar battles in the corridors of some of the world's great workplaces, though no doubt with greater sophistication.

GET THE MACHINE GUNS READY

For example: When a power-sharing arrangement was under consideration after the merger that created Citigroup in 1998, a senior executive named Jamie Dimon remarked, "Get the machine guns ready. Co-CEOs will set up an obsession with who's winning, and factions will discredit each other and destroy careers."

Dimon was entirely correct. The co-CEOs, Sandy Weill from the Travelers insurance company and John Reed from Citicorp, clashed from the start, and it took less than two years for Weill to drive Reed into exile, a contest described in journalist Monica Langley's biography of Weill, *Tearing Down the Walls.*

Meanwhile, Dimon, who was Weill's protégé, unwisely chose not to promote Weill's daughter, Jessica Bibliowicz, a senior executive with Citigroup's Smith Barney brokerage. He also criticized her in front of others. Dimon was a banker, not a psychologist. Even so, it was a remarkably elementary mistake to imagine that, when it came to the boss's daughter, job performance mattered more than family ties.

Bibliowicz soon left the company. Dimon subsequently became embroiled in his own power-sharing struggle during the early months at Citigroup. Weill, still simmering over his daughter's departure, crossed over factional lines to side with Dimon's Citicorp rival, discrediting Dimon, if not quite destroying his career. Dimon went off into exile, becoming head of Chicago's Bank One.

These fights typically have little to do with making a better product—the perfect cinnamon bun—or building a better company. Those are the pretexts on which we fight for the things that matter in our primate hearts. It is our nature to want to work with other people, to covet the sense of affiliation with them. But we also want to have rank, dominance, social status, power over them. Call it ridiculous, but one of the essential issues with which we struggle from birth to death is the "I'm in charge"/"No, you're not" dichotomy. It matters to us even when we know better, even when we resort to the ultimate one-upmanship of opting out: "Touché, dude. Your donuts rule."

And it matters much more when the players choose not to opt out. Jamie Dimon, for instance, proved himself a dynamic leader at Bank One. *Fortune* magazine depicted him boisterously taking the stage at a company pep rally to the Steppenwolf tune "Born to Be Wild." On stage, he shouted, "What do I think of our competitors? I hate them. I want them to bleed." Then in 2004, Bank One merged with J. P. Morgan to create a company that could rival Citigroup as the world's dominant financial institution. The deal brought Dimon back to New York as CEO-designate. And it immediately became evident exactly which competitor Dimon really wanted to see bleed: Sandy Weill and Citigroup. "If he doesn't understand that," Dimon told the *New York Times* with a predatory laugh, "he will."

W. Edwards Deming, the management consultant, used to lament the endless round of faction fighting and one-upmanship within companies. A histrionic speaker with a booming voice, Deming liked to milk a sad story about a five-year-old girl going to a Halloween party, a happy event until (ominous drumroll) "a prize was offered for the best costume." According to journalist Art Kleiner, who followed Deming on the lecture circuit, his voice would drop to a whisper as he recounted how the little girl, and all the others who had not won, went home in tears. "Deming wrapped up the story, week after week, by crying out in a hoarse, anguished moan: '*Why does somebody always have to win?*' "

Apparently, no one ever stood up to give Deming the answer:

It's because we are primates.

RISING TO C LEVEL

Samuel Johnson, that large, learned ape, almost certainly understated the human instinct for forming hierarchies when he said, "No two men can be half an hour together but one shall acquire an evident superiority over the other." It doesn't take a half hour. In a Stanford University study, groups of male college freshmen put in a room and given a problem to solve needed less than fifteen minutes to sort themselves into hierarchies. Children from the age of five onward also spontaneously form social hierarchies, with or without a prize for best costume, and though the criteria may seem vague—who's toughest, or coolest, or most popular—they readily agree on where everyone ranks. These hierarchies prove remarkably stable as the children grow older.

As adults, we jockey for a better position in the workplace hierarchy, of course. But even the appearance of a better position seems to gratify some craving for which bread alone does not suffice. In a British survey, 70 percent of office workers said they would forgo a pay raise in exchange for a better job title—"data storage specialist" instead of "filing clerk," "catering supervisor" instead of "tea lady." During the dot-com euphoria of the 1990s, some startups consisted of a CEO, a dozen vice presidents, and a "director of first impressions," otherwise called "the receptionist." One American retail chain, taking advantage of this built-in status hunger, used the title "manager in training" as a way to get glorified sales clerks to work extra hours at the cash register without overtime pay.

Were they out of their minds? If so, it was, and remains, epidemic. You may consider yourself above this sort of status hunger, and so did I when I was a young newspaper reporter honing my jaded attitude in Elizabeth, New Jersey. But I called myself Union County bureau chief, though I personally constituted my paper's entire Union County bureau, with an annual payroll topping $12,000. Such dubious tokens of status should presumably become steadily less important as we grow older and make our way up the economic ladder. But our appetite for them seems only to increase.

At the top of most American companies—what consultants call "C level"—everybody wants to be chief, though they are generally careful to disguise it with an abbreviation. A company's top management will thus frequently include a CMO (chief marketing officer), a CFO (chief financial officer), a CIO (chief information officer), a COO (chief operating officer, or hatchet man), and a CEO (the demigod). The story used to be told about a Deutsche Bank chairman that when he got to heaven (it was a fairy tale, evidently) he found it in economic shambles and proposed a financial restructuring. The plan proved impractical, because God balked at being deputy chairman of the board.

All this is donut dominance writ large. The appetite for rank and status is so intense that we seem to re-create the highly codified hierarchies of the playground wherever we go for the rest of our lives. We fret endlessly about who's got the best office, the biggest budget, the hottest BlackBerry, and other minutely calibrated workplace distinctions.

"In the world of Hollywood, you have to think like Hollywood," one major movie executive recently declared. Then he rattled off the local hierarchy in precise detail: "There's the major motion picture from a big studio, then there's the independent successful film, then big television, then the good art film, then the bad one, then bad television, then schlock movies, and then celebrity boxing, the bottom . . . rung." This comment was occasioned by the arrest of a forgotten television star, Robert Blake, for allegedly murdering his wife, and the studio exec went on to say: "These people aren't even at celebrity boxing status. Why should we care?"

It reminded me of one time I was out with a biologist in Botswana and she was detailing the whole hierarchy of who grooms whom among the baboons she was studying. Then she turned to a dismal character off to the side. "And then there's Bob," she said. "Nobody grooms Bob."

Hierarchy is ubiquitous among humans and other social animals because we are born wanting it. Every child starts out in the hierarchy of the family, nosing eagerly into the bosom of parental authority, which feeds and comforts us, and protects us from the terrors of the outside world. In time, we begin to shift our sense of dependence onto other adults, also seemingly larger than life, and even eventually onto people our own age. (The larger-than-life attraction persists: In one study, half the CEOs in Fortune 500 companies stood six feet or taller.)

When we begin to work, if we have the good fortune, we apprentice ourselves to a mentor who nurtures us like a parent within the hierarchy of the workplace. We learn to compete with other people at about our own level, much as we used to compete with siblings, and perhaps in time we manage to assert dominance over them. For social animals like us, being caught in a harsh social hierarchy, or being stuck at the bottom of the heap, may sometimes seem bad. But the really frightening thing, so frightening that we can linger immobilized for years within a destructive hierarchy, is when we find ourselves cast out by our old social network. The phone stops ringing. People look the other way in the halls. It dawns on us that we are alone.

MACHIAVELLIAN MONKEYS

The tendency to form hierarchies isn't something we have learned through trial and error. We haven't reasoned our way there. On the contrary, we often try to reason our way out of it: "Am I really giving up my weekend just to make the boss happy? Do I really need to stay at the office past my kids' bedtime on the chance that it will get me a promotion sometime next year?" We certainly haven't come to accept hierarchy because some bright soul in strategic planning told us that it was the path to improved profitability.

Hierarchy is in our genes.

Monkeys and apes are exquisitely attentive to questions of rank, and this is where the obsession with hierarchy took hold of our minds. In Botswana, where I got to wander in the middle of a troop of baboons and begin to know them on something like a first-name basis, this was the single most startling point of recognition. The baboons fretted endlessly over who got the best spot in the jackalberry tree, or who got second crack at the palm nuts. They were acutely conscious of where every other monkey stood in their troop. They knew one another as individuals, as members of families, and as participants in social networks bound by ties of friendship or political alliance. When somebody attacked a young male, for instance, the other monkeys didn't just pay attention to the conflict; they also looked at the victim's brother, to see if he would come knuckle-racing to the rescue. Biologists regard this kind of "social intelligence" as the defining characteristic of social primates like us.

The other term biologists often use is "Machiavellian intelligence." The baboons I was following vied constantly with other individuals at their level. When the alpha male drove the beta away from a choice palm tree, for instance, the beta soon recouped his pride with a quick, sharp glower to drive off the male immediately below him (who passed it on in turn). The ground was littered with enough palm nuts to keep everyone contentedly gnawing into next fiscal year. But the baboons seemed most covetous of the palm nut their neighbor had just picked up. Along with grooming and other more amiable forms of interaction, an obsession with the "I'm in charge/No you're not" dichotomy ate up their days. And why not? Alpha baboons, like alphas in human groups, tend to get their choice of food, shelter, and sexual opportunities.

The maneuvering for status apparently only becomes more intense as we move closer to humans in the primate evolutionary tree. Harvard biologist Richard Wrangham, who has studied chimpanzees in Uganda's Kibale Forest and in Tanzania, writes: "[A] male chimpanzee in his prime organizes his whole life around issues of rank. His attempts to achieve and then maintain alpha status are cunning, persistent, energetic, and time-consuming. They affect whom he travels with, whom he grooms, where he glances, how often he scratches, where he goes, and what time he gets up in the morning. (Nervous alpha males get up early, and often wake others with their overeager charging displays.) And all these behaviors come not from a drive to be violent for its own sake, but from a set of emotions that, when people show them, are labeled 'pride' or, more negatively, 'arrogance.' "

Now read that paragraph again, substituting "ambitious manager" for "male chimpanzee" and "nervous executives" for "nervous alpha males." It could run almost unedited in any number of business profiles in the *Wall Street Journal* or the *Financial Times.*

One reason management consultants have generally refrained from pointing out the remarkable and instructive similarities in hierarchical behavior between humans and other primates is to spare the feelings of the executives who hire them. Even without dragging apes into the equation, social dominance has always been a delicate subject. If you doubt this, try telling your coworkers how deeply you yearn to lead them. At least some will hear this as "get you under my thumb."

Dominance behavior has also often been neglected in the business setting because it is a relatively new subject for scientific study. Terms such as "pecking order" and "alpha male" date back less than a century, and "alpha female" is of considerably more modern vintage. The existence of a "dominance drive" was first proposed in the 1930s by a young primatologist named Abraham Maslow, who was studying captive monkeys and chimpanzees. If Maslow's name sounds familiar, that's because he later made his reputation as one of the great corporate management theorists of the twentieth century, writing about human motivation. Maslow's "hierarchy of needs," an explanation of what motivates people on the job, has become conventional wisdom among managers.

Maslow also coined the term "self-esteem" as a way to talk about the feeling of dominance without needlessly reminding people of the under-my-thumb subtext. "My primate research is the foundation upon which everything rests," he once wrote. But perhaps prudently, his business writings never talked about the ape and monkey studies that helped shape his ideas about human hierarchies.

So can we really learn anything about our own dominance behaviors by looking at other animals? As always, it depends on how we look. Early researchers often worked with animals housed in separate cages rather than in social groups. This made rugged individualism more or less mandatory and helped perpetuate the top-down, command-and-control idea of dominance. In the field, researchers often studied baboons, largely because they spend much of their time on the ground rather than in the treetops, so it was easier to see what was going on. But baboons are about as crude and brutal relative to chimpanzees as chimpanzees are to humans. So that research also tended to reinforce the autocratic corporate-warrior brand of social dominance.

Because wild chimpanzees spend much of their time in the treetops, biologists didn't even realize until 1968 that they lived in stable groups, or that these groups had complex social lives, much less that their social lives might contain useful lessons for ours. Toshisada Nishida, a Japanese primatologist working in Tanzania's Mahale Mountains, was the first to describe the group behavior of wild chimpanzees. Nishida, Jane Goodall, Richard Wrangham, and other field researchers soon came to recognize individual chimps by appearance and family history, and thus began to piece together the soap opera struggle for social dominance among wild chimpanzees. Their research, along with Frans de Waal's work on captive chimps, has shifted the dominance debate away from the individual and onto the social context, where relationships, reciprocity, and continuity matter, much as they do for us.

HIERARCHY WORKS

Whether we happen to be watching baboons, or chimpanzees, or humans in a corporate hierarchy, the W. Edwards Deming question inevitably comes to mind: Why does somebody always have to win? And why do we care so badly? The conventional Darwinian explanation is that status matters because, sooner or later, it leads to sex. But this must fall a little flat for high-powered executives putting in one hundred-hour workweeks. Their appetite for status is so intense that it frequently obliterates the appetite for sex.

Why has the drive for status, dominance, and social hierarchy become such a powerful force in our lives? The answer is that donut dominance isn't, after all, ridiculous. The distinctions for which we plot and scheme in the workplace aren't frivolous. Hierarchy works, on many levels.

▪ *The hierarchy benefits even bottom-feeders.* Out on the African savanna, where our simian ancestors lived and evolved for millions of years, outcasts had an unfortunate tendency to get picked off by lions or neighboring troops. So even Bob, the neglected schlub at the bottom of the heap, tended to do better in a hierarchy than he would on his own. The other members of the group served as extra eyes to watch out for predators, or, failing that, as extra bodies for Bob to hide behind.

Individuals who played by the rules of the group—that is, the ones who made suitable displays of deference to social superiors and otherwise cultivated useful social alliances within the group—generally got to stick around. If they were particularly savvy about working their way up in the hierarchy (or clever about exploiting those who did), they might also get to pass on their more sociable genes to future generations. Thus the Darwinian forces of natural selection and sexual selection ensured that a penchant for hierarchy got encoded in our genome.

Subordinates also fared better in the hierarchy because the alpha typically took the lead in territorial defense and in the fight for resources. This remains true in modern-day hierarchies, though we mostly notice when the alpha neglects this role entirely. At one publishing company, for instance, the staff loved their division boss, who ran a decent workplace and seldom had an unkind word for anyone.

Unfortunately, he also avoided controversy of any kind, and because he loathed office politics, he never bothered to appoint some scrappy lieutenant to fight on his division's behalf. The company was moving to new quarters. Other divisions, where the managers had a keener focus on rank and privilege, ended up with fewer cubicles and more real offices. They got doors! They got windows! The placid manager's division meanwhile got partitions. In fact, his demoralized employees were obliged to inventory their existing furniture so that the other divisions could cherry-pick the best stuff for their new offices. Sometimes, though we may not like to think it, it can be better to live under a difficult, demanding alpha.

▪ *The hierarchy helps ensure domestic tranquility.* Hierarchies often seem like the cause of endless internal squabbling, in the manner of Sandy Weill and John Reed, or Jamie Dimon and Jessica Bibliowicz. The group can end up paying more attention to issues of rank than to making a profit. Lou Gerstner complained that when he arrived at IBM in the early 1990s, Kremlinology was a fine art, and the first chart in any presentation, regardless of subject, "depicted the internal organization, including a box showing where the speaker fit on the chart (quite close to the CEO most of the time)."

But fights merely catch our attention. Animal studies suggest that a relatively settled hierarchy actually means *less* squabbling, not more. Figuring out who's in charge in the first place can certainly be bloody and painful. Fighting is five times more likely to occur in chimp groups, for instance, when rank is uncertain. The same is true in corporations after a merger. But once a hierarchy gets established, ambitious subordinates are unlikely to mount an overt challenge. It's simply too risky. Having a strong leader also makes any group more effective at coordinating action toward some clearly stated goal. We may resent the hierarchy. But we also find comfort, safety, and purpose within its boundaries. Likewise, when a flock of chickens has a stable pecking order, the hens fight less and lay more eggs.

Biologists increasingly view social hierarchy not as a tool for encouraging and rewarding aggressive behaviors but as a way to control aggression and steer it into socially acceptable behaviors. The classic case in the natural world involves African elephants orphaned by ivory poachers in the 1980s. Young males growing up without the calming (or intimidating) presence of adult males turned into juvenile delinquents. They experienced early and unusually intense musth, a recurring period of hormonal change when males engage in aggressive behaviors and compete for sexual opportunities.

At a game park in South Africa, orphans went on rampages, killing large numbers of white rhinos. Biologists hit on the solution of introducing mature males. After encountering these physical and social superiors, the juvenile males quickly began to behave in a more civilized fashion. The presence of elders actually suppressed the signs of musth in the juveniles, and the rampages ended. It doesn't take too big a stretch to see the analogy between the delinquent elephants and the misbehavior of twenty-somethings in the dot-coms of the late 1990s. Entrepreneurial companies often require the arrival of older, steadier hands to survive.

The tranquilizing effects of the hierarchy work the other way, too. The respect and status granted by subordinates help to civilize the alpha. Even the most august CEO cannot lead a hierarchy for long if he lacks a solid base of support among his subordinates. Smart bosses thus go out of their way to win over underlings, and they depend on a few trusted lieutenants to let them know when they are failing to do so. Arrogant bosses may get away with bad behavior for a while. But they eventually end up like Disney CEO Michael Eisner, forced out of his chairmanship by disgruntled shareholders, or like one of de Waal's alpha chimps, who behaved so badly he was besieged by a coalition of angry subordinates and left "high up in a tree, alone, panic-stricken and screaming."

▪ *Higher-ranking individuals set standards and serve as models.* Sometimes, to be sure, they are bad models. For instance, in *Pipe Dreams,* his book about the Enron debacle, Robert Bryce calculated that this insanely profligate energy company routinely spent $45,000 per weekend using its Falcon 900 jet to ferry an executive, Lou Pai, back and forth between his suburban Houston home and his Colorado ranch. Pai was "a genius at trading," but his division lost sacks of money. Moreover, the precious executive time that he saved using the corporate jet often got spent reading the newspaper in the office or stepping out for lunch at local strip joints, also on the expense account. Such behavior by Enron executives served as the model for a culture of kleptomania. Hence the spectacle of white-collar looters wheeling their "personal items" out the door after the company's collapse, using their Aeron chairs as shopping carts.

Models of more sober leadership are less fun to read about, but they are more common in companies that actually manage to survive and return a profit to shareholders. For instance, Richard Kinder left Enron in 1996 to form his own gas transmission company, Kinder Morgan, and made it a success on a tightwad business model. "Our goal is getting everybody to think, 'How would I spend the money if it was my money?' " Kinder recently told the *Wall Street Journal.* Kinder owns 20 percent of the company, so it *is,* to a large extent, his money.

Nonetheless: "Most CEOs try to sell that message to their people. But if you are out in the field trying to get them enthused about watching costs and they see you get into a chauffeur-driven limousine back to the Kinder Morgan jet, they are going to just laugh you off. If they knew you stayed at Red Roof Inns, it begins to have some impact." Kinder Morgan has no corporate jet, and Kinder says he flies coach. He also draws a $1-a-year paycheck, or 93 cents after tax withholding. But the frugal model has made him a billionaire, and many of his employees, who also own stock, are on their way to becoming rich.

JetBlue founder David Neeleman practices a similar style of leadership by unpretentious example: All JetBlue employees are required to clean the planes on which they fly, even as passengers on vacation. Neeleman joins in picking up discarded newspapers, used airsickness bags, and other passenger detritus. So do the pilots.

- *The hierarchy motivates people.* The tendency in recent years has been to focus on the ways hierarchy can destroy motivation, especially when the kleptocrats at the top grab a disproportionate share of the rewards. But no sane shareholder would suggest that getting rid of the hierarchy is the way to solve the problem.

The unequal distribution of rank and privilege, when properly managed, is one of the things that keep people showing up for work every morning. Little tokens of status—the company car, the new job title, the deference of subordinates—encourage us to progress through the ranks. "Privileges still exert a powerful, positive motivating force, especially on junior members of a team," says General Motors vice chairman Robert A. Lutz. "When I was an aspiring leader, I *wanted* a reserved parking space (and worked hard to merit one)."

Likewise, Jack Welch quit his first job at General Electric after one year not because his $1,000 raise was trivial on objective grounds, but because it was the same raise given to two other young managers who had joined the company at the same time. Welch relented only after a smart boss intervened and elevated him by an extra $2,000. It was the normal status rivalry of a young chimp eyeing his neighbors' palm nuts. Welch also counted the ceiling tiles in his office, to be sure his office space increased in tandem with his status in the company, and faster than the other guys'. It is tempting to laugh at that spirit of one-upmanship. But it ultimately helped bring him to the top, where he applied a consciously unegalitarian distribution of incentives and responsibilities to drive his entire company to extraordinary levels of achievement.

EGALITARIAN TALL TALES

Welch claimed to hate hierarchies. But everybody claims to hate hierarchies, as if in our hearts we are all fiercely egalitarian, all earnestly meritocratic. Even "Chainsaw" Al Dunlap boasted in his business memoir that there was "no forbidding hierarchy around me." (Less than a page later he also described himself tearing through one hapless company "like Godzilla through Tokyo.")

In fact, a long line of scholarly thinking treats hierarchy as an unnatural state, a modern imposition on free-spirited, brother-and-sisterly human nature. Anthropologists describe the traditional tribes to which we belonged through almost all of human history as fiercely egalitarian. Hunters strutting homeward with their prize warthog carcass didn't slice off the best bits for themselves. They shared it equitably with other members of the tribe, or if they failed to do so, somebody eventually sliced the best bits off them. Leadership also typically seemed to shift depending on which tribe member was best qualified for a particular job. The best tracker moved ahead on the hunt, for instance. The keeper of tribal wisdom took the lead in rituals. Individuals relished their autonomy. They took it badly if some would-be *jefe* tried to bully them into his way of doing things.

To skeptics, the idea that traditional tribes were egalitarian may sound suspiciously like the idea that they were savage innocents. Anthropologists holding up the egalitarian ideal generally don't bother explaining how we could have been so hierarchical through tens of millions of years of primate evolution, then so egalitarian for perhaps a hundred thousand years of tribal life, then even more fiercely hierarchical beginning twelve thousand years ago, with the invention of agriculture and the rise of human wealth. One writer goes so far as to describe our social evolution as a U-turn.

This is not a maneuver customarily taught in the Darwinian Driving School, but it may well come close to the truth about the nature of human hierarchies. Evidence increasingly suggests that primates have shifted back and forth between egalitarian and authoritarian styles of social dominance repeatedly throughout their history. It may depend on the circumstances: In war, the command-and-control model, with barked orders instantly obeyed, seems to work more effectively. In peace and prosperity, when the threat is not so imminent and unhappy individuals can easily go elsewhere, a more egalitarian style usually makes more sense.

The dominance style of a primate group may also vary depending on the personalities of high-ranking individuals. It's common among chimpanzees to shift from leadership by a relatively benign alpha, such as the animal Jane Goodall named Freud in Tanzania's Gombe National Park, to an unmitigated brute such as his brother Frodo, who displaced Freud and went on to terrorize his troop. (He also terrorized Goodall and the cartoonist Gary Larson, among others, thumping them into bruised, angry submission.) The dominance style in a particular group may also depend on the personalities of the subordinates. If they band together, they can sometimes oust a despot, or maneuver to ensure that a less tyrannical individual becomes the alpha. But even the most egalitarian primate group still has a hierarchy.

The mistake the anthropologists make (along with the rest of us, much of the time) is to regard egalitarian moments as evidence of an egalitarian society. Like every primate social group, hunter-gatherers undoubtedly had their alpha males and females, which is to say they had their hierarchies, however invisible they may have seemed to the intrusive eyes of anthropologists. Most traditional tribes were no doubt highly cooperative. *But sharing some resources doesn't necessarily make any group egalitarian.* On the contrary, generous behavior has always been one of the most effective ways to establish social dominance, for alpha chimps and corporate chief executives alike.

- Ntologi, one of Toshisada Nishida's chimps in the Mahale Mountains, routinely used gifts of food to cultivate useful allies. He was a fierce hunter himself and would also sometimes snatch food from other successful hunters, then share it in a feeding cluster with females, younger males, and influential elders. You could call this philanthropic and egalitarian. But it was certainly smart leadership behavior, and it enabled Ntologi to maintain his alpha status almost continuously for sixteen years, until his death in 1995.
- Similarly, among corporate chieftains, Nike chairman Phil Knight was undoubtedly being cooperative and philanthropic when he promised $30 million to help his alma mater, the University of Oregon, renovate a football stadium. But he reneged on his promise when the university joined a group that monitors Third World sweatshops such as the ones where Nike manufactures sportswear. University president Dave Frohnmayer, whose enthusiasm for defending academic integrity was apparently limited, promptly severed ties with the sweatshop-monitoring group. The university continued to display its submissive rump to Nike for the next seventeen months, with its stadium project on hold, before Frohnmayer and his jubilant head football coach were able to get back up on their feet and announce the resumption of Knight's "magnitude."

For Knight, as for Ntologi, the egalitarian gesture of sharing wealth was in fact a tool for maintaining authoritarian social dominance. Cases like this, and there are many of them, suggest that we should probably drop the phrase "giving back" from the language. Philanthropy is sometimes just a nicer form of taking.

For tribal hunters, likewise, bringing home the bacon and sharing it with the tribe was a way to win over potential allies and wow local females. Alpha males in traditional tribes got a disproportionate share of matings, arguably the most important resource. Alphas also commanded respect and a degree of deference to their political opinions. When hunter-gatherer tribes settled down to an agricultural way of life, with the opportunity to store food and hoard wealth, these muted hierarchies blossomed. Even the Hopi Indians of the American Southwest, long celebrated by anthropologists for their egalitarian way of life, had a clear hierarchy based on the size and quality of a family's landholdings, and also on their family mythology. In a drought, the chiefly clans got to stay put and survive. Underlings starved or went elsewhere, in a sort of tribal downsizing, often to be slaughtered by neighboring tribes.

"There are no egalitarian societies," a modern anthropologist concedes. "Nor . . . are there any simple societies," merely "egalitarian contexts, or scenes, or situations."

EGALITARIAN EMPLOYERS

Like ancient tribes, modern companies often hold up the egalitarian ideal. Intel, for example, likes to celebrate its open culture, with no reserved parking spaces, no executive dining rooms, small cubicles for all employees, and everyone working together on a first-name basis. But the first name Andy still puts a quaver in people's voices.

As CEO, Andy Grove often "worried about what he was not getting for his money," according to one former Intel executive, and he used to require company workers arriving after 8 a.m. to sign a "late list." He also once sent out a companywide notice, immediately dubbed "the Scrooge memo," advising employees to work a full day on Christmas Eve. (He did not specify if he meant the ten- or twelve-hour day that was common for many workers, or if an eight-hour day would suffice.) Another high-ranking Intel executive once defined fear as the moment you are going through your in-box and discover an overlooked Andygram, the nickname for Grove's often abrasive e-mails, and it's marked "AR," for "action required," and it's already one day old, and right then the phone rings, and you know it's Andy. *"You just know."*

Hierarchies are the inevitable human condition. They may be brutally confrontational and yet brilliantly successful, as at Intel. They may be brutally confrontational and scandalously bad, as at the Boston mutual fund giant Putnam Investments (where longtime CEO Lawrence J. Lasser also used to browbeat employees in terse "Lassergrams"). They may often be relatively democratic and nurturing. But for all their egalitarian ideals, even undeniably gentle bosses usually find some subtle way to assert their status. Sometimes it's innocuous. At one Connecticut company known for its nurturing climate, all employees have standard e-mail addresses on the lines of JaneDoe@widget.com. The founder's address, on the other hand, is IAMME@widget.com. (Possibly short for IAMWHOAM@widget.com.)

These unexpected glimpses of naked power can also sometimes be surprisingly raw. At Cisco, CEO John T. Chambers has always cultivated a folksy, friendly image. His office is an open twelve-by-twelve-foot workstation, and he avoids meeting people at his desk lest this put him in a "position of power." He prefers to use a round conference table, he says, to emphasize that "we're in this together."

But critics have depicted Chambers as unwilling to hire senior managers who might threaten his own status. When a longtime executive vice president made the mistake of presenting himself as heir apparent in a *Fortune* magazine profile a few years ago, Chambers promptly arranged for the vice president and his wife, who was head of corporate communications, to pursue exciting new career opportunities outside Cisco. At the announcement of his departure, the doomed vice president appeared before his staff with the company's top executives lined up behind him. In *Cisco Unauthorized,* by Jeffrey S. Young, one witness described the scene as "looking like a Mafia hit team rub-out."

If these sound like egalitarian companies, you probably also believe that Wal-Mart workers are actually "associates," not retail peasants, and that people in Communist societies really stood side by side as comrades. (No doubt it was that precious fraternal moment just before they pitched forward into their collective graves.)

COVERT HIERARCHIES

We seem to have evolved to deny the existence of social dominance even as we pursue it with all our hearts. The result is that this central fact of our working lives is often invisible not just to outside observers but even to ourselves. A recent article in the *Harvard Business Review* suggested just how intangible a hierarchy can be: "Where does power lie? Who's parking next to whom in the parking lot? Who's the first to speak after the CEO in meetings? An executive can pay a high price for missing such hierarchical cues. It is one of the costs that the denizens of hierarchies must pay for the rewards they receive . . . pressure to remain constantly on the qui vive to avoid inadvertently stepping on the wrong toes."

At a Michigan company, a rumor went around that a senior vice president had been asked to find another job. But at a Christmas party that year at the most exclusive country club in the area, a subordinate who was clearly on the qui vive noticed that the senior vice president had his nameplate on the locker next door to that of the CEO/chairman. "I deduced (correctly so far) that the rumor was just someone's wishful thinking. Can you imagine having to sit next to someone that you just fired every time you went out to have fun playing golf?" (The only dubious idea here is that playing golf is about having fun.)

For human beings, the obsession with social dominance is generally invisible, unconscious, and yet continuous, like breathing in and breathing out. A hierarchy can be frustrating; it can cause people to leave their jobs and seek someplace where the Deming question—"Why does somebody always have to win?"—does not apply. But no such place exists.

When primatologist Terry Maple got his first teaching job after graduate school, for instance, he thought a university faculty would be largely free from the bothersome claptrap of corporate hierarchy. Professors, after all, enjoy the security of lifetime tenure. Then he attended his first department faculty meeting, where no more than twenty people, freethinkers all, would be spread out around an auditorium seating two hundred. Maple, the overeager novice, was first to arrive and found a seat in the third row. "Another faculty member came in, whom I barely knew," Maple recalls. "He kind of stood around, looking at his shoes, looking a little sheepish. It was an awkward situation. I was in the third seat and he was in the aisle. I said, 'Well, what's going on, Tom? And he said, 'You're sitting in my seat.' And I started laughing. I got up and moved one seat over, and he said, 'Now you're sitting in Professor Smith's seat,' and I said, 'Well, that's too bad. Professor Smith is going to have to sit somewhere else.' "

Maple managed to keep his job, and it occurred to him as a primatologist to take notes on the phenomenon, if only as a way to make faculty meetings less dull. Over the next few years, he documented "rigid, repeatable seating preference even after moving to a new meeting room." It was an invisible hierarchy. The only significant disruption occurred when the first women faculty joined the department. In that dim era before political correctness reined in baser primate urges, the alpha predictably permitted these young women to sit close to him, a familiarity that would have been intolerable for male newcomers finding their place in the hierarchy. (At least he didn't sniff the air and snort.)

Maple knew that other species also have behavioral devices for invisibly staking out their turf and establishing a stable position relative to potential rivals around them. Wolves use scent marking as a spacing mechanism. Birds do it with song. These small markers of territory and dominance have a straightforward function in the animal world. They declare, "I live here, back off," and thus minimize the possibility of face-to-face conflict with neighbors. The unhappy alternative would be for everyone to spend all their time fighting, and animals are too smart for that. It takes too much energy, diverts them from the main task of putting food on the table, and exposes them to a considerable risk of having their throats ripped out.

It's the same for us on the job, and the ways we conform to the rules of the invisible hierarchy go well beyond seating order. As we shall see in later chapters, we unwittingly declare our dominance or deference not just with body language or facial expression, but every time we open our mouths, and also when we have the good sense to keep them shut.

So what's the bottom line? How much hierarchy or dominance does a company need to get the best results from workers? It depends on the circumstances. Baboons and Fortune 500 companies can and sometimes do make a U-turn in their dominance style, when circumstances or personalities dictate. The important point is that, in all circumstances, workers are comfortable with clear lines of hierarchy. They need people above them who are willing to make decisions (even the wrong decisions) promptly, rather than letting them lie around, as a financial analyst once put it, "like three-legged horses that no one wants to shoot." They just don't like to be reminded of the hierarchy too openly, or too often.

Maple later became CEO of ZooAtlanta, a job that required him to meet regularly with many of the most important figures in Atlanta's booming business community. He noticed that the same sort of seating order preference was a standard feature of these meetings, and that it functioned much like the dominance order in a group of baboons. He came to think of boardroom behavior as "Baboonology 101," an insight he was generally careful not to share with his fellow primates. "I consider it a compliment to be compared to the highly successful, astute baboon," Maple remarks. "But I fear that my business colleagues may not share my enthusiasm."

Most of them seemed never to see—and in truth never wanted to see—the hierarchy in which they nonetheless instinctively took their seats.

Clownfish are brightly colored little coral reef fish living in small groups among the tentacles of anemones. The tentacles are poisonous to outsiders. But they provide shelter for perhaps a half dozen clownfish, which thrive on microscopic organisms drifting through and on feces from the anemone. Going out into open water entails a high risk of being picked off by predators.

So the only hope for a newly hatched clownfish is to be taken on as a "junior partner" by a clownfish hierarchy in a nearby anemone. The newcomer strives to look small and unthreatening so the resident clownfish won't chase him off. This works so well that clownfish turn it into a career strategy. At every stage in the hierarchy, individuals carefully maintain their size at just 80 percent of the size of the individual immediately above them. By never rising above their station, never threatening the boss, clownfish get to live (and, okay, eat poo) for another day.

6

TOOTH AND CLAW
How We Wage Dominance Contests on the Job

Diplomacy is the art of saying "Nice doggie" until you can find a rock.

—Will Rogers

When he was a high-ranking executive at the BBC in the 1990s, Will Wyatt frequently dealt with another top executive, Patricia Hodgson, and he recalls that she had her personal assistants "play the game of making sure that you came onto the telephone before she did. 'Shall we go through now?' her [personal assistant] would ask yours, the convention being, and one would say common courtesy, that both put their bosses through together. But no. Only yours would press the connect button so that you had to speak to the PA before you were allowed to speak to Patricia herself. It irritated other PAs, who were made to feel junior. I just smiled."

Well, Wyatt smiled, but also managed to get this dig onto page 7 of his memoir *The Fun Factory: A Life in the BBC.*

Little contests like that are a daily, even hourly, part of our working lives. Two people make eye contact across a room and one of them holds the stare. Bingo, we have a dominance contest. The one who averts his eyes first surrenders. Little acts of territoriality, little displays of status, serve to put the other guy on edge, at a disadvantage. Language infinitely expands the subtlety and complexity of these exchanges. An alpha chimp has no more than about twenty vocalizations with which he can let his rivals know that he is a nasty, rotten son of a bitch who means to get his way.

A corporate executive saying essentially the same thing can draw on a vocabulary of sixty thousand to one hundred thousand words (Can you say, "I want this on my desk before you leave tonight"? Can you say "outsource"? Can you say "skill rebalancing"?), along with a rich repertoire of body language and facial expressions. This allows the imprecations, the admonitions, the bending, spindling, and mutilating, the hints that one is seeking candidates for downsizing, to come across in a whispered phrase, or even in a mere tone of voice. Once when they were both executives at American Express, Sandy Weill turned to Lou Gerstner at a meeting of high-ranking colleagues and, planting his thick finger on a footnote in a report, demanded, "Why has the time to answer a customer call gone from thirty-eight seconds to forty-five seconds?" It was a fair question and more or less friendly, the ordinary give-and-take of daily business. But anybody who ever worked with Weill would recognize that it was also an attempt to make Gerstner lose face.

A human dominance struggle can take place even on terms that are arguably flattering. For example, a Latin American boss was visiting a facility in Puerto Rico run by one of his subordinates, an American. "You look very sexy today," he told her, in front of a team of visiting American researchers. From his perspective, it was a gracious compliment. From hers, it was an attempt to puff himself up by folding her into his harem. In another time or culture, she might have acknowledged this act of dominance with a submissive smile, or a modest glance downward. Instead she offered him an icy smile. Then she turned to the researchers and said, "Forgive my boss if you're in culture shock now." It was a graceful way of calling him a macho boob, while also acknowledging him as "my boss." The researchers may hardly have noticed that a thrust for a certain kind of dominance had just been parried, or that a hapless male stood neatly deflated before their eyes, with everyone still smiling.

We are seldom conscious of these little exchanges as actual contests, even when our faces are flush with the emotions they produce. We are even less likely to notice the little acts of deference and reassurance that we pay to those in power as a means of averting a dominance contest in the first place. In the modern workplace, the business of dominance and submission is usually most effective when it is least obvious. Almost all of the aggression gets channeled into symbolic forms. That's the main function of job titles, big offices, power clothes (as the phrase "Armani armor" suggests), the seating order at meetings, important friendships, ample expense accounts, and other signs of power or status.

It's why, for instance, the executive suite used by top management is almost always demarcated, regardless of the company or the architecture, by an eerie, cloisterlike silence. "If you walk in talking in a normal voice, rather than sotto voce," says one consultant, "or if you laugh, God forbid, all the administrative assistants look up sharply as if you have broken the *omertà.*" The effect is to induce soft words and humble postures in petitioners granted an audience there, and this is exactly as intended.

At least one U.S. president has noted, with evident glee, the way visitors rehearsing verbal assaults outside his door are reduced, on entering the Oval Office, to making inane remarks like, "You're looking good, Mr. President." The alpha who is secure in his status never actually needs to declare his dominance. His trappings do it for him.

People thus sometimes fool themselves into believing that social dominance isn't an issue at all. We make the mistake of thinking that humans are engaged in dominance contests only when they raise their voices, and that animals exert dominance only through open combat. The truth is merely that *we pay more attention* to open combat because it's more exciting. It makes better television footage if a bull elephant spills a rival's blood with his tusks. It makes better scuttlebutt around the watercooler when a junior associate gets reamed in front of everyone. But even bull elephants are generally smart enough to establish who's in charge without needless bloodshed. Or as a Swahili proverb puts it, "When elephants fight, only the grass gets hurt."

RITUALIZATION

This is the first lesson about social dominance: It rarely involves overt physical force. Frans de Waal once watched an alpha chimp, Nikkie, chase his rival Luit around their compound at the zoo. Chimpanzee males have five times the upper body strength of a college football player, and roughly the same sense of decorum. So other chimps leaped for safety as the two combatants kicked up clouds of dust. You could hear their screaming at the far end of the zoo. Nikkie and Luit ended up in the bare branches of a dead oak tree, safely separated, panting as they slowly calmed down. But for all the uproar, *neither of them had ever actually laid a hand on the other.*

Rival baboons likewise generally avoid combat. Instead, they sit in the treetops screaming "wah-hoo" at each other until the one who shouts longer and louder finally prevails. They get their way by showing they might just become dangerous one of these days. Biologists refer to these displays of aggressive intent as *ritualization.* They enable the animals to trade ritualized signals instead of bites or blows.

Ritualized threats exist for the very businesslike reason that they are cost-effective. In the course of evolution, surly individuals who indulged in violent aggression often experienced short, unproductive careers. Even if they somehow managed not to get killed, they diverted too much of their energy from feeding and finding mates to have much success at the vital business of passing on their genes to future generations. Animals that got ahead by bluffing, blustering, and other forms of ritualized aggression, on the other hand, lived to dandle their grandchildren.

Violent coercive dominance behavior is rare in the human workplace for the same reason—because it's so dangerous. Even shoving a coworker is liable to land you in the hospital or the courthouse, and out of a job. We are also acutely aware that the use of force can work both ways, not just from the top down, but from the bottom up, as a tool used by subordinates to back off an unwelcome manager.

The other problem with force as an instrument of social dominance is that it reveals weakness, not strength. For example, some managerial types like to walk around fondling a baseball bat or a golf club, to make the point that they literally carry a big stick. An executive producer supposedly affected this style at a television news show in Chicago some years ago. Maybe he was just a Cubs fan. But one day a nervous young intern joined the staff. There was nowhere for him to sit, so someone kindly set up a folding table and a chair in a corner of the newsroom.

"Who the fuck put this here?" the producer demanded as he was making his rounds a little later. Then he smashed the bat down on the table while the nervous young man was sitting at it. Recounting the story years afterward, the intern was still visibly shaken by this display. He knew that walking around with a baseball bat was an homage to that other Chicago managerial genius, Al Capone. On May 7, 1929, Capone threw a lavish dinner party for his associates, with numerous toasts to the guests of honor, gunmen Albert Anselmi and John Scalise. Late in the evening, Capone's festive mood suddenly vanished. He had the two men gagged and lashed to their chairs and revealed that they had been planning to betray him. Then he took out a baseball bat and methodically beat them about the head and the shoulders, leaving them to be finished off by a bodyguard with a bullet to the brain.

Colleagues are naturally intimidated by a show of force. But they also know that confident alphas get what they want with the merest lifting of an eyebrow. So they interpret lashing out as the first fatal sign of vulnerability. As the financial fraud at Enron was beginning to unravel in April 2001, for example, CEO Jeff Skilling held a crucial conference call with major stock market analysts and institutional investors. A hedge fund manager who had made a large bet that Enron was about to tank asked several critical questions. Skilling's answers were evasive, so the hedge fund manager upped the dominance challenge. "You're the only financial institution," he said, "that can't produce a balance sheet or a cash flow statement with their earnings" before a conference call.

"Well, thank you very much," Skilling shot back. "We appreciate that. *Asshole.*"

A moment of stunned silence followed. Stock analysts who had been muttering "buy, buy" like well-tended sheep suddenly snapped awake to the idea that they might be on their way to the slaughterhouse. Enron stock plunged.

Likewise, Capone's active career ended with his arrest on an unrelated charge just ten days after his infamous batting practice. And when I tried to track down the bat-wielding Chicago news producer, the somewhat vindictive story I heard was that he had found a new career selling Amway products and spreading the message of Jews for Jesus. I eventually located him by phone. He said that he had been a line producer at the station in question, not an executive producer, and he denied ever having attacked a table with a baseball bat. He had in fact left television and was a "normal person" now, as he put it, with "various business interests."

In the animal world, ritualized displays come in all sizes and shapes: The frilled lizard lifts itself up, fans out the cowl around its throat, and hisses with mouth agape. The spotted skunk does a sort of handstand, showing off its black-and-white warning coloration and the big, threatening tuft of its scent-spraying tail. The mantis shrimp simply holds up its spread claws as if to say, "Oh, yeah? Just try me."

Humans also practice a vast repertoire of ritualized dominance behaviors. You do it when you stamp your foot and glower at your dog. (Your dog slinks off with tail between its legs, a ritualized way of saying, "I'm sorry, don't hurt me.") Likewise, when the boss leans forward at a meeting, or points emphatically, or pounds the table, or lets his voice slide down to a growl, he is announcing to all that he is on the brink of going feral. Chimps, baboons, and alpha humans can sometimes win a dominance struggle with nothing more overt than a sharp glance, the ritualized prelude to a charge.

In one experiment, researchers reared infant monkeys in isolation, then exposed them to photographs of angry adult monkeys. The infants had never seen an angry face before and thus had had no opportunity to learn what it means. But it still made them panic. For young monkeys and probably for humans, too, an angry face is a natural warning signal, and a fight-or-flight response to this signal seems to be coded into the genes.

Companies sometimes consciously design ritualized threat displays into their products. Chrysler marketing consultant Clotaire Rapaille boasted, for instance, that the Dodge Durango SUV was designed to make other drivers feel they were sharing the road with a savage jungle cat. "A strong animal has a big jaw, that's why we put big fenders," he told Keith Bradsher, author of the SUV exposé *High and Mighty*.

The Durango was made for drivers with a heightened "reptilian instinct for survival," according to Rapaille, who added, "My theory is the reptilian always wins. The reptilian says, 'If there's a crash, I want the other guy to die.' Of course, I can't say that aloud." So his savage-jungle-cat SUV says it for him.

Clearly, ritualized threats aren't always subtle. King Charles II not only removed Oliver Cromwell's rebellious head from his body but kept it posted over Westminster for twenty years as a ritualized way of saying, "Regicides need not apply." During his considerably less glorious reign as CEO at Sunbeam, Al Dunlap livened up his office with images of eagles and lions, a decorative technique for unnerving visitors by eliciting an unconscious predator response. Dunlap also planned his office at the company's new headquarters in Florida with a spare room for his dogs and another for his bodyguards. Complaints, anyone?

In the same minatory spirit, Enron employees referred to their corporate headquarters as the "Death Star" and to chief executive Jeff Skilling as "Darth Vader." When chief financial officer Andrew Fastow went over to the Dark Side, creating the little side deals that let him skim off millions of dollars in illicit profits, he named them after predators—Timberwolf, Bobcat, Osprey, Egret, and Harrier, among others. It would no doubt have been more accurate to name them after parasites—Tapeworm, Maggot, and Bloodsucker come to mind—but it wouldn't have been nearly so effective at subtly discouraging scrutiny. At Kmart at one point, the vice president for Human Resources, of all things, was a former helicopter pilot nicknamed "the Rottweiler." He was part of a leadership team known inside the company as "the Frat Boys." Among other threat displays, they liked to throw darts in the conference room at pictures of Wal-Mart and Target executives. Then they went bankrupt.

THE DEMONSTRATION TANTRUM

Journalist Owen Edwards experienced the effectiveness of the ritualized threat in a form he labeled "the demonstration tantrum." In the late 1970s, he was the managing editor of *Cosmopolitan* magazine and came to know editor Helen Gurley Brown as "all in all a very hardworking, fair-minded, demanding boss . . . who never indulged in public humiliation or executions in the village square." But one day, Edwards heard a loud thumping on the wall between his office and Brown's. When he went to investigate, editorial assistants "had already gathered at the door of the boss's office and were staring in, mouths agape. In a systematic frenzy, Helen was picking up everything on her desk and hurling it full force against the wall. This went on until everything that could be thrown had been thrown, by which time the entire staff had gathered to watch. It turned out that Helen's rare explosion had come about because the Dictaphone system had malfunctioned, and the twenty or so letters she had recorded before leaving the office late the night before had vanished into the ether. So none of us was the object of her incandescent wrath. But no one who witnessed the scene ever, *ever* wanted to piss her off again."

In the same spirit—that is, abject submission—an industrial engineer named Ron Godbey recalls how the latest of many reorganizations at Hamilton Sundstrand, an aerospace company, put him under a new boss, who was short and fat, with "tiny little arms" and a nasty personality. Jabba the Hutt, as he was inevitably nicknamed, arranged to meet his new subordinates one-on-one in his office. As Godbey was waiting in the reception area for his appointment, a coworker emerged "with no color in his face and . . . the look of someone who had just undergone electric shock therapy." Godbey gathered up a folder of "major accomplishments" and went in for his session. As he arrived in front of the massive desk, Jabba looked up: "Godbey, right?" he asked. "My impression of you, Godbey, is that you are a nice guy. I *hate* nice guys. Now get the hell out of here." It was like opening a kitchen cabinet and finding yourself staring down the throat of a rabid schnauzer. Godbey spent the next couple of years lying low, as Jabba plainly intended by his ritualized display. "It was a time of massive layoffs. I just did my job and hoped it would still be there the next day."

BEING PROSOCIAL

Jabba the Hutt was clearly not a model boss. Or rather, he was a boss on the model of U.S. secretary of defense Donald Rumsfeld, who kept a bronze plaque on his desk at the Pentagon quoting Teddy Roosevelt: "Aggressive fighting for the right is the noblest sport the world affords." (This from a former naval aviator who never saw action.) When he was

U.S. secretary of state, Colin Powell, a combat veteran from Vietnam, suggested another approach to social dominance with the quote from Thucydides he displayed on his desk: "Of all the manifestations of power, restraint impresses men the most."

Many psychologists now argue, alongside Thucydides, that "aggressive fighting" isn't the most effective way for humans to wield social dominance. You don't necessarily get to the top by overtly aggressive behavior. In many cultures, jabbing your hand in a rival's face or barking at the secretary will merely get you labeled a social misfit. You can build your leadership style, or your product line, on the "reptilian mind." But we are primates, not reptiles. Our brains and our repertoire of social behaviors have expanded enormously over the course of our evolution, and it is a shame to ignore all that. Even apes often practice the style of leadership known as prosocial dominance, as Ntologi demonstrated with his food sharing, and Yeroen with his political maneuvering.

Prosocial (as opposed to antisocial) dominance is about gaining power by deploying the gentle forces of compromise and persuasion. It's about knowing how to get what you want with a self-effacing joke. It's about finding ways to help subordinates instead of browbeating them. People who become bosses often lament the continual need to cajole rather than command. You have to motivate workers who are mistrustful and resistant to change. You have to avoid letting yourself be riled by minor acts of subversion.

An American boss, for instance, heads a staff of ten in a Latin American country. Her male administrative assistant behaves deferentially in private but likes to assert his independence by flouting her authority in front of other males. His favorite put-down is *"No-no-no-no, no me entiendes,"* meaning, "You just don't understand what I'm saying," or more pointedly, "Your Spanish isn't good enough." She is trying to get him to switch to "Let me try to be a little more clear" or "Let me say what I mean another way," but the more deferential tone is precisely what he doesn't want. Even worse, one of the other male staffers has picked up the patronizing phrase.

At such moments, her job feels to her like a continual contest for power, and this is entirely correct. If being the boss often means being nice, showing tact, and practicing restraint, it is still ultimately about who's in charge. At such a moment, it's her job to take the subordinate aside and explain as plainly and unemotionally as possible, "Here's the standard of behavior I expect from you. You need to know that if you act like that again, I will fire you."

In other words, being positive and prosocial does not mean being a pushover. Well-meaning managers often go out of their way to lavish praise on subordinates, even when they perform poorly, in the belief that it's a way of building confidence (not to mention racking up points for their five-to-one ratio of positive to negative interactions). But subordinates are not stupid. They know when you are patronizing them. So false praise can come across as a put-down (and, *egad,* add points to the negative side of the ratio). For instance, minority students often perform worse, not better, when teachers set out to "build self-esteem" by coddling them. Either they get deluded into thinking they're doing high-quality work, setting themselves up for eventual disappointment, or they see through the praise: *"I get it, she thinks I can't do the work because I'm black."* They excel, on the other hand, when a teacher expects a high standard of performance and provides the individual help and attention to convince students that they can get there.

Prosocial dominance is often about getting the best work from people by finding ways to minimize the emotions that inevitably attend any dominant-subordinate relationship. For instance, when social workers try to persuade clients to stop using drugs or having unsafe sex, it's implicitly a dominance relationship: *What I'm telling you is right, and what you're doing is wrong.* The clients often hear this as a threat to their sense of self-worth. They respond with what social psychologists call a "defensive bias," discounting the source of the information or twisting it into something that actually reinforces the original misguided behavior. Likewise, when a boss tells the sales staff that they aren't hitting their numbers and asks them to try a new tactic, they're likely to roll their eyes and attribute the sales drop to market conditions or some other factor outside their control.

But psychologists have found that if you first give people a reason to feel good about themselves—for instance, by getting them to write down a few things they did really well over the past quarter—then they will be much less threatened and more open-minded when you ask them to consider the new tactic. "The commonsense approach," says Yale psychologist Geoffrey Cohen, "is to affirm someone in the same domain as the threat: 'Your sales last year were great. This year they're substandard.' " But this approach is likely to backfire. Boosting an individual's sense of self-worth to mitigate a threat works best, according to Cohen, when it's about something in a totally separate domain. You might say, "That speech you gave at the Hong Kong meeting was phenomenal. People are still talking about it." Then you have a foundation to move on to an unthreatening discussion of what's going wrong with this year's sales.

Such emotional considerations form the unconscious subtext for our working lives, and they are deeply rooted in the biology of social dominance. Psychologists define power as the ability to modify other individuals' lives either by controlling the flow of resources (such as food, money, economic or decision-making opportunities, friendship, and support), or by administering punishments (such as physical harm, job termination, verbal abuse, or ostracism). The workplace is quintessentially about both forms of power. And even though we do nothing more overt than sit in a room and chat, the ups and downs of power have a peculiar knack for setting our hairs on end and pricking every molecule of our being to life. Social dominance influences every nuance of our behavior, even when we think it is the farthest thing from our minds.

JUST KIDDING

Even contests that are entirely ritualized or strictly verbal have physiological consequences. Allan Mazur, an engineer and sociologist at Syracuse University, suggests that rival humans "compete often for status in fairly well-defined contests, each trying to 'outstress' the other through actions or words . . . The level of stress that is experienced may be high, as when the contest takes the form of aggressive combat, or very low, as when the competition takes the form of polite conversation, in which case the participants may barely be aware that they are in a contest."

In one experiment, researchers paired an undergraduate with strong feelings about abortion against a stooge pretending to believe the opposite. A neutral introductory conversation produced no signs of stress by a standard measure, thumb blood volume. But when the conversation turned to the hot topic, both parties reacted like animals in a dominance contest, the blood rushing from their thumbs to the major muscle groups, the first step in a fight-or-flight response, as each struggled to outstress the other. Mazur writes: "Finally, one individual 'surrenders,' accepting the lower rank and in so doing alleviating the stress of the competition."

The same sort of responses occur during dominance contests in the workplace. For instance, a testing laboratory was doing environmental analysis on samples from an office tower that had suffered a fire. Some of the sample swipes contained a substance thought to cause birth defects. "Our whole lab staff was women of childbearing age, many of them married, and we were not equipped to be working with level-three toxins," says the safety person. She walked into a meeting to address the issue and realized with dismay that she was facing a posse—the president of the company, the head of environmental testing, the lab director, and the head of engineering. The moment of truth came when the president said, "So you don't think our lab is equipped to handle these samples?"

"Yes, that's correct," she replied.

"And that's your prerogative."

"Yes."

"And my prerogative is that I can fire you."

"Yes, that's your prerogative."

"You're fired."

It was a surreal moment, hearing those words most bosses scrupulously avoid. ("We have to let you go," they say, as if you are straining to boot yourself out the door.) No one had raised his voice or displayed the slightest emotion. The safety person quietly gathered up her things and left the office, then ran to the women's room and fell apart. Ten minutes later, the president's voice came over the loudspeaker, addressing her by name: "I know you're in the building and I'd like you to come back to my office."

She pulled herself together and slouched back to find the same cast of characters waiting. "I was just kidding," the president said, smiling wanly. The others had apparently advised him on the biohazards of firing a whistle-blower. Soon after, the company took all the questionable samples and sent them away. Then it gave the safety person a huge raise and suggested that she keep her mouth shut, now that the problem had been resolved. A month later, she found a new job, and a month or two after that, the samples came back. Dominance contest over.

THE PHYSIOLOGY OF DOMINANCE CONTESTS

The first thing that happens during such dominance contests is that your body reacts to the threat before you are even conscious of it. Your senses are constantly scanning for danger, and they automatically pick up a colleague's threatening body language or facial expression. The boss brushes by and growls, "We need to talk. *Right now.*" Within a tenth of a second after the first sign of a threat—that grimace, that angry, impatient movement, that tone of voice—signals start to flash around the amygdala and other first-response areas of the brain. It takes much longer, a half second or so, for the plodding verbal part of your brain to catch up to what he just said and think, *Uh-oh.* But your body is already in gear.

Indeed, the beauty of some ritualized displays, such as the savage-cat appearance of the Dodge Durango, is that they may never actually rise to the level of consciousness. But people still instinctively veer out of the way. The adrenal glands pump out the alarm hormone epinephrine (also called adrenaline) to prime muscles and organ systems for action. At the same time, ten thousand little franchises of your sympathetic nervous system launch jolts of norepinephrine to stomach, esophagus, mucous membranes of the nose and palate, small and large intestine, and other muscles and organs, notably including the rectum, now suddenly clenched.

Your pulse quickens. Your arteries contract. Your blood pressure rises. Your stomach goes taut. Your face gets hot. You may experience a state of piloerection—that is, your hairs rise, as when a lion puffs himself up for combat (though the effect in naked apes like us is sadly diminished). If the discussion wells up into a confrontation, you may unconsciously knot your fist.

TESTOSTERONE FICTION

A dominance contest, or even the mere prospect of such a contest, also causes a surge in testosterone, particularly in men. It doesn't much matter if it's a battle to the death with mace and broadax or a debate over the location of this year's company picnic. Conventional wisdom says good meetings go bad and companies go awry when men get their testosterone up, sending the discussion spinning off into eyeball-to-eyeball overtime. Meanwhile, the women exchange baffled glances. Did these guys really just go toe-to-toe about whether the company newsletter should run a photo of Snoogy, the senior vice president's dog? Like men trying to comprehend the emotional influence of estrogen, women just don't get it when it comes to testosterone.

Unfortunately, men don't get it either. Our ideas about the role of testosterone in workplace disputes are almost entirely false. This masculinizing hormone, found in men at seven times the concentration in women, is still a code word for aggression and violence. But biologists have found no evidence that testosterone causes conflict. In fact, having an elevated baseline testosterone level tends to be characteristic of confident, socially adept males who are least likely to use overt aggression to achieve dominance. Bullies and other sociopathic types typically display *lower* testosterone levels.

Moreover, the temporary surge in testosterone during a dispute is too slow to have much effect on the outcome, according to Stanford University physiologist Robert Sapolsky, who studies stress in humans and baboons. It takes about an hour after a conflict begins for testosterone levels to peak, and the only relatively rapid change is an increase in the rate of metabolism of certain muscles after about forty-five minutes. Increased testosterone seems to be an *effect* of conflict, not a cause. The winner really feels it an hour or so after exiting the battlefield, or the conference room, when he goes striding around the hallways with his chin up, bearing confident, and testosterone spiking.

Testosterone seems to matter in workplace dominance contests mainly because of this so-called winner effect. Those who have won once are also more likely to win the next time around. They walk onto the battlefield with steady gaze and confident posture, and this body language can be a self-fulfilling prophecy of continued success, if only because of the powerfully disheartening effect on rivals. The tantalizing possibility is that ambitious individuals may be able to manipulate their own testosterone levels by the more or less natural means of defeating their peers—in a squash match with somebody you're not quite sure you can beat, a pickup basketball game against the kids from the mailroom, or even a chess match with your top strategist. These little contests can boost your testosterone level and help you head into the boardroom afterward with chin up and momentum at your back—that is, assuming you win. The temptation is to schedule a match with a patsy, the klutz from accounts payable, say. But the mind-body dynamic is sophisticated and not all that easy to fool, so winning a sure thing may not give you the winner effect you're hoping for.

The other intriguing discovery about testosterone is that even people who merely side with one of the parties in a dominance contest seem to enjoy a trickle-down winner effect. During the 1994 World Cup games, for instance, Brazil beat Italy. Brazilian fans who watched the game on television experienced a rise in testosterone levels. Italian fans went into a testosterone slump.

This suggests that it might be possible for effective leaders to spread the winner effect among their subordinates—for instance, by ensuring that assembly line workers invest themselves in the contest when Airbus takes a big contract away from Boeing, or when Matsushita gets a product to market ahead of archrival Sony. Conversely, the repeated experience of defeat turns some animals into "trained losers." They give up before a contest begins. Leaders who blame and belittle their subordinates or otherwise make their social dominance too explicit may thus risk producing an organization of trained losers.

CORTISOL AND CONFUSION

In both men and women, dominance contests also cause a surge in cortisol, an adrenal hormone. It happens almost instantly, unlike the testosterone surge. Fifteen minutes after the start of a conflict, cortisol levels have doubled and, because of peculiarities in biochemistry, this produces a tenfold increase in effect. Despite cortisol's bad reputation as "the stress hormone," this is a good thing. In a brawl, we need plenty of cortisol to keep up the blood pressure, boost blood glucose concentration, and otherwise make us strong enough to withstand the rigors of combat.

In a crisis, dominant individuals tend to show a sharp spike in cortisol, which then quickly subsides to a relatively low resting level. That is, they have the stuff to recognize real threats and rise to meet them. Subordinates, on the other hand, often suffer a relatively high level of cortisol even at rest and don't mount much of an increase in cortisol during a crisis. "Their signal-to-noise ratio isn't great," says Sapolsky.

In one study, researchers divided boot camp recruits according to social rank in their group. Then they subjected them to stress tests, which included delivering a speech and performing mental arithmetic in front of an audience. The salivary cortisol levels of dominant individuals shot up, to 14 nanomoles per liter. Subordinate individuals experienced only a slight cortisol uptick, to 2.9 nmol/L.

An elevated cortisol level makes us uncomfortable. In fact, it can do much worse than that over the long term. Chronically elevated cortisol, from a stressful job featuring lots of conflict, abuse, and uncertainty, can lead to high blood pressure, muscle weakness, diabetes, osteoporosis, suppression of immune function, an increase in infections, infertility, weight loss in the arms and legs, and a weight gain where you least want it, in your tubby gut. (Maybe this is why Dilbert looks the way he does.) Over the long term, high levels of cortisol can also cause cell death in a part of the brain called the hippocampus. In extreme cases, such as post-traumatic stress disorder and depression, the hippocampus begins to atrophy, resulting in loss of memory, impaired cognitive function, and inability to regulate emotional display in a normal manner.

Despite the extraordinary toll on the individual, tactics that elicit elevated cortisol levels are standard at some companies. Physical restraint, though unorthodox, occurs surprisingly often—for instance to minimum-wage night-shift workers locked in at Wal-Mart. Or to Arthur Andersen auditors put in a meeting room at Enron and told that they could not leave till they signed a letter approving a dubious $270 million tax credit scheme.

Shouting is also commonplace, and Gap Inc. recently estimated that up to a quarter of its factories in Asia rely on psychological coercion or verbal abuse. The retail chain made the revelation as part of an unusually frank effort to improve factory conditions. Many other companies tolerate or even encourage verbal abuse, though the victim's biochemical response to this type of dominance display is much the same as in a physical attack. Studies suggest that verbal abuse can be even more emotionally destructive than a beating.

At one Detroit automotive supplier a few years ago, Meredith Johnson joined the company as a senior manager during a time of extreme pressure to satisfy rapid growth in sales. Her job description by itself might have made her a dreaded figure. She was a "change agent," often stereotyped in corporate nightmares as an outsider with a clipboard and a stopwatch (or, more likely these days, a personal digital assistant), intent on finding faster, more efficient ways to get jobs done. Johnson (not her real name) actually managed to be worse than the stereotype. She made the people under her produce major reports on unrealistic deadlines. She frequently called meetings for 8 a.m. Monday and then did not bother to darken the company doorstep until 10 a.m. When subordinates displeased her, she made a point of announcing in front of coworkers that if they did not shape up she would e-mail human resources and get them fired. She actually fired three of them over a period of months. She also talked with HR about people outside her department whom she deemed "vulnerable."

"The fear factor was causing anxiety, sleeplessness, and other physical symptoms of stress," says another senior executive at the company. "One of her subordinates, who thought he was immune because he sucked up to her, suffered a heart attack. People felt they had no control over their jobs or their destiny."

Johnson's intimidating behavior went unchecked for two years. People submitted not just out of fear but because Johnson had made a point from the start of talking to the male CEO as if they were best pals. He never saw her dark side. So any complaint would have seemed like whining or unprovoked mutiny. Even when Johnson's subordinates finally mustered the collective will to take their situation to HR, the company still gave her months to find a new job.

The damage she did lingered long afterward. People who had worked under her, many of them ten- and twenty-year employees, "no longer trusted the company that they had given their hearts and lives to when sales were down. They will never have the same loyalty they had before." It would be pleasant to report that Johnson ended up unemployed and suffering from, say, chronic, uncontrollable flatulence. But in fact she lives on and continues to practice coercive dominance behaviors in the management of another Detroit automotive company.

People who work for such bosses generally try to resolve conflicts as quickly as possible, to bring down their cortisol level and regain their inner composure. One way they do that is to surrender. They avert their eyes. They back out of the boss's office nodding. After having been "outstressed," as Allan Mazur puts it, "submission is the price one pays in order to relieve one's feelings of discomfort." Even in companies that go out of their way to nurture employees, deference to those higher up in the hierarchy becomes standard operating procedure. "As associates, nodding our heads in agreement was an involuntary reflex," write John Rolfe and Peter Troob in their book *Monkey Business,* about their time as investment banking underlings at Donaldson, Lufkin & Jenrette. "We'd been taught to always concur. Conflict was confusing."

GO FORTH AND PROSPER

One of the most celebrated bids for social dominance in modern organizational history involved a Tanzanian chimpanzee named Mike, who did not at first seem destined for the fast track. Then he discovered the noisemaking potential of the empty kerosene cans at Jane Goodall's research station. Even better, he found that he could scare the wits out of animals who outranked him by running at them with an empty can clattering ahead of him through the undergrowth. By chimp standards, this was ritualized aggression of a high order. He could also make rivals scatter by merely putting on a fierce face and banging two cans together. Within four months, Mike passed over his dumbfounded superiors and became the alpha male in his troop.

Bluster, bravado, or the big noise can also sometimes be an effective tool for getting ahead in the human workplace. For example, as a young executive at Lucent, Carly Fiorina was once assigned to subdue the unruly staff of a recent acquisition. The other company had a macho culture, and its employees disdained their new masters as geeky and effete. So what did Lucent do? It risked reinforcing all their stereotypes by sending out a woman in a pantsuit for the big meeting. In his book *Perfect Enough,* business writer George Anders describes how Fiorina gave her spiel, starting gently, then gradually becoming more macho than her macho audience: "We at Lucent think you guys are a bunch of cowboys," she said finally. "You probably think we're a bunch of wusses. Well, I think it's important that we get to know each other." Then she stepped out from behind the podium to reveal an absurdly large male bulge at her crotch. (Her husband's gym socks, actually, but who knew?)

"Our balls," she remarked, "are as big as anyone's."

The effect was much like Mike rolling his empty kerosene cans, with startled subordinates scattering into the treetops crying "Waaa! Waaa!" It was an act of sheer bravado. The shrewdly calculated jibe at both herself and her audience instantly moved them past their preconceptions about one another. Fiorina later went on to become CEO at Hewlett-Packard and for a time, before her confrontational style brought her down, one of the world's most powerful business leaders.

Both Mike the chimp and Carly Fiorina make case studies for an intriguing new theory about who gets power, how they get it, and how it affects them and the people around them once they've got it. And no, it doesn't necessarily take big balls or empty kerosene cans. What Mike and Fiorina alike possessed was a precious commodity, which in the framework of the "approach/inhibition theory of power" might be termed the fast-forward personality. The approach/inhibition theory was proposed in 2003 by psychologists Dacher Keltner of the University of California at Berkeley, Deborah Gruenfeld at the Stanford Graduate School of Business, and Cameron Anderson of New York University's Stern School of Business.

It begins by asserting what ought to be obvious: Power is central in human relationships. Or as Bertrand Russell once wrote, "The fundamental concept in social science is Power, in the same sense that Energy is the fundamental concept in physics." Ignoring it is like certain Victorians hoping sex would go away if they just stopped thinking about it.

Beyond that, the theory makes an audacious attempt to put together all the elements of power and social dominance in a single comprehensive framework, encompassing phenomena as diverse as ritualized aggression, the winner effect, the sloppy table manners of corporate moguls, the tendency of women to be more alert to office politics than men, and even the varying level of complexity in opinions issued by the U.S. Supreme Court.

Keltner and his coauthors view power as essentially a question of whether to approach and be expansive or to withdraw and be inhibited. It's the watering hole scenario again, though the authors do not elaborate on the evolutionary background. The approach/inhibition theory argues that powerful people are quicker to detect and approach opportunities for rewards, including food, shelter, attention, sex, and money.

Having less power, on the other hand, leads to greater activation of the "behavioral inhibition system," or what other authors call negativity bias, meaning heightened anxiety, social constraint, and attention to threats and punishments. The authors skirt the issue of whether individuals are born one way or the other, with a predisposition to be powerful or subordinate. They are far more intrigued by the question of how getting or losing power changes the way anybody behaves. Power itself, they suggest, "tips the balance" in a person's mind, from inhibition to approach.

ME WANT COOKIE

In a way, the approach/inhibition theory got its start in the bewildering tendency of powerful people to behave in socially inappropriate ways. Deborah Gruenfeld had worked in the magazine business and sometimes attended meetings at *Rolling Stone* magazine where founder Jann Wenner ate raw onions and slugged vodka from the bottle in midconversation, without offering similar refreshments to his guests. A friend of Cameron Anderson's had been through an oral exam for his doctorate at which one faculty member not only picked his ear wax but held it up in the light and studied it lovingly. And everyone had heard about President Lyndon B. Johnson's penchant for receiving cabinet members, newspaper reporters, and other underlings while seated on the toilet.

Indeed, conducting personal hygiene in front of subordinates, generally taken to be a weird foible of your boss or mine, turns out instead to be a rich tradition among alpha types. In the approach/inhibition scheme of things, it's one way bosses reveal how disinhibited they are. A current top executive at the *New York Post,* for instance, used to step aside during editorial meetings earlier in his career to urinate in a corner sink. During meetings at another company, the boss sometimes flosses his teeth. A magazine publisher receives subordinates and her pedicurist simultaneously. And at a prominent political lobbying firm in New England, a partner sometimes farts loudly and without comment during meetings with staff. It isn't a medical condition; he knows not to fart while importuning public officials for a favor worth $2 million. Nor is it a sign that he and his staff can be completely at ease with one another. Everyone in the room understands that his farting does not constitute permission for them to fart, too. It is the subordinate's place to endure the scent marking of the alpha, not the other way around.

Keltner and a colleague, Andrew Ward, had previously devised a whimsical test of the hypothesis that power makes people less sensitive to those around them. They set up an experiment with groups of three subjects, randomly giving one person in each group the role of handing out assignments to the other two. After leaving the three for a half hour to work through "a long and rather tedious list of social issues," the experimenter came in with a plate containing five cookies. Not only was the individual who had been given the power role more likely to take a second cookie, he was also more likely to eat it with his mouth open, spew crumbs on partners, and get cookie detritus on his face and on the table.

The cookie experiment—perhaps "cookie monster experiment" would be more apt—was never actually published. But it helped lead Keltner and his coauthors to their approach/inhibition theory. It fit neatly with a large body of previous studies associating power with traits the authors now identified as part of the behavioral approach system: Individuals with power (even randomly assigned power) tend to talk more, interrupt others, speak out of turn, and engage more readily in conflict. They use more expansive body language and smile less (subordinates specialize in smiles of submission or appeasement). They're more likely to enter the social space of others, stand too close, initiate physical contact, and flirt in less inhibited ways.

ONWARD CRUSTACEAN SOLDIERS

The usefulness of the approach/inhibition theory as a way to understand power is only now being tested by other researchers. Assuming it holds up to scrutiny, the beauty of the idea is that it seems to provide a coherent explanation of the behavior of creatures from flatworms to corporate CEOs. It also appears to make sense in terms of the biochemistry underlying those behaviors.

For example, serotonin is a neurotransmitter; it functions to ease the passage of signals between nerve cells. An increase in serotonin in the brain seems to make people more relaxed, self-assured, and socially assertive. Low serotonin, on the other hand, tends to make people more anxious, quarrelsome, and impulsive. Drugs such as Prozac treat depression by manipulating the serotonin level. Research has also demonstrated that serotonin plays an important role in social dominance. Among vervet monkeys, a rise in social rank leads to a corresponding increase in serotonin levels. Loss of rank causes serotonin levels to drop. Officers in college fraternities likewise tend to have higher serotonin levels than do their rank-and-file brothers—not, apparently, because they were born that way but because power has tipped the balance of their internal biochemistry.

From the approach/inhibition perspective, the intriguing thing about serotonin is that it apparently got its start early in animal evolution as a sort of "red light, green light" biochemical. In primitive organisms such as crustaceans, its function is simple: A rise in serotonin causes an increase in motor activity. A drop in serotonin inhibits movement.

This makes sense for the animal's survival. In circumstances where the food supply is abundant, moving forward means being able to go after food and consume it. On the other hand, if predators are lurking nearby, immobility and inhibition make sense, to avoid being noticed. Thus serotonin apparently started out as a simple approach-or-inhibit mechanism and only later, with a little rejiggering, became a key factor in the fast-forward personalities of high-powered people and in the inhibited and constrained behaviors of their subordinates.

POWER HAZARD AHEAD

Even if we accept that flatworms and vervet monkeys can teach us about power and social dominance behavior in humans, does the approach/inhibition theory really matter? Does it help people behave more intelligently in the workplace? Armies of $3,000-a-day consultants have not yet leaped up to carry the message into the corporate workplace. But they will, because people who choose to be ignorant about the innate tendencies associated with power risk being undone by them. In fact, they are highly likely to be undone by them, because ignorance or obliviousness about what's going on around them seems to be one of the primary tendencies in question.

"We assume," Keltner and his coauthors write, "that power activates the behavioral approach system without conscious awareness of its effects and, in fact, that those with power may actually be less cognizant of others." That is, the boss really is off in his own little world, much as we have always suspected, and this is in large part the nature of being a boss. As power increases, firing up the behavioral approach system and shutting down behavioral inhibition, people become less attentive to those around them. They often place too much faith in their own ability to shape events.

This exaggerated sense of power is why bosses frequently take credit for work actually performed by subordinates, who bite their tongues and bide their time. It's why bosses react to negative events with anger (meaning something ought to have been done), whereas individuals given low-status positions react with guilt and sadness (meaning, whatever it was, we failed to do it).

People with power also "construe social events in more automatic fashion," and often in less complex ways. This may happen partly because they have too many other things to think about. But it's also because they have less reason to worry about the consequences of their actions.

To test the idea that people with power often function on autopilot, Deborah Gruenfeld looked at opinions issued to support majority decisions by the U.S. Supreme Court over a forty-year period. Each of the opinions was written by one justice on behalf of a coalition of other justices, and Gruenfeld took the size of the coalition as a gauge of the author's power. The larger the coalition, she found, the simpler the opinion tended to be, and the less constrained. The smaller the coalition (and presumably the more active the author's behavioral inhibition system), the more deliberative and complex the resulting opinion. That is, without apparently being aware of it, judges dedicated to "equal justice under law" instinctively conformed to their realpolitik sense of where the power lay.

If these are natural tendencies associated with power and rank, how does being aware of them help us behave more intelligently? How does a smart leader use these tendencies to win dominance contests where necessary, but also avoid bigfooting people or otherwise antagonizing them when there is no need? The approach/inhibition theory suggests several useful caveats:

- Having a leader with a heightened sense of personal power—a leader with the nerve to shake off impediments and cry, "Damn the torpedoes! Full speed ahead!"—clearly helps, in many circumstances. It helps, at least, when the torpedoes don't happen to be armed with nuclear warheads. But it can also be catastrophic, as when AT&T boss C. Michael Armstrong went on a shopping spree for cable companies in the boom years of the late 1990s—an investment he was forced to sell off just three years later at a loss of $50 billion. To put this debacle in perspective, the entire company was valued at just $15 billion when SBC Communications proposed acquiring AT&T early in 2005. An awareness of approach/inhibition behaviors should alert powerful people to be more vigilant about controlling their own tendency to ignore contradictory facts and to favor information that fits their stereotypes or preconceptions.
- One way to maintain vigilance is to cultivate a close relationship

with relatively low-ranking individuals, because people with less power typically see the world more clearly than do their bosses. The logic of this is straightforward. While powerful people are paying attention to the potential rewards, disempowered people are paying attention to the likely costs. They have to stay alert to what's going on around them because they are more vulnerable to threats of all kinds, from layoffs to nasty assignments nobody else wants to handle. (Physiologist Robert Sapolsky has immortalized this idea in proverbial form: "If you want to know if the elephant at the zoo has a stomachache, don't ask the veterinarian, ask the cage cleaner.") A smart boss thus learns, at the very least, to respect and depend on an executive assistant.

Though it is hardly ever stated out loud, one of the assistant's key functions is to serve as extra eyes and ears, swapping gossip with other assistants and picking up little hints of coming events that might otherwise blindside her boss. In scientific studies, women often seem to have better social antennae than men. This isn't necessarily because being female makes you smarter or more sensitive to the nuances of human behavior. Given power, women have amply demonstrated that they can be as stupid and insensitive as men. But most women still lack power. So being more attentive to others is a survival technique they have developed to help wend their way through potentially threatening social environments.

- Because subordinates have superior social antennae, it can also be smart to let people with less power conduct important negotiations. Keltner and his coauthors point to studies showing that high-powered sorts "tend to be less aware of their opponents' underlying interests." Low-powered negotiators are not only more aware but also more likely to find solutions that work for both sides.

- Powerful people sometimes forget that their hard-charging personalities elicit a deep sense of threat in subordinates. This becomes dangerous when subordinates are too fearful to speak up and stop the boss from charging into a brick wall. Bosses also often combine intimidation with an even more dangerous tendency to impose stereotypes on subordinates. It's not necessarily because they're bigots, but because of the autopilot thinking associated with power. In one study, even college undergraduates tended, as they became more powerful and began hiring people, to view job applicants less as individuals and more as stereotypes.

This combination of social threat and stereotyping can be toxic. In another study, African Americans performed as well as European Americans on the Graduate Record Exams and other tests, until the experiment prompted them to think about their race. Women did as well as men on math tests, except when told that the test was designed to reveal gender differences.

■ Powerful people need to be aware how quickly their skewed view of the world can lead them into the worst sorts of public scandal. The same driving personality that caused John Rigas to found a small cable company and build it up into a huge public corporation also encouraged the delusion that Adelphia Communications was still "his" company and that everybody else would get out of his way. Rigas and one of his sons were convicted of stealing more than $2 billion from the company, which went bankrupt.

Power also leads naturally to sexual misbehavior. The "simple idea of power," Keltner and his coauthors write, causes people to entertain increased thoughts and feelings about sex, "especially in those individuals prone to disinhibited, inappropriate sexual behavior." They develop a sense of entitlement. For instance, a BBC chairman admonished for putting dalliances on the expense account was simply annoyed. "Fuck off," he replied, and then with wonderful insouciance, he added. "I'm an old man. I'm going to die soon and if the BBC doesn't like it, too bad."

But times have changed, and understanding the propensities associated with power can provide useful perspective: A male boss may recognize that it's the power relationship itself that's causing him to regard a female subordinate as sexually available. Thus he may (but, okay, probably won't) spare himself the garish blunder of taking it as a romantic overture when she is merely making the friendly, deferential displays subordinates routinely pay their bosses.

Or maybe the woman is genuinely attracted to her boss. An understanding of the power relationship may give him pause to reflect on whether it's his shiny scalp and his trembling, protuberant belly that she finds so hot-hot-hot. Or is it just his corner office? With luck, this thought may be enough (but, again, probably not) to persuade him to take his propensities outside the workplace, where the perils of sexual harassment do not loom so large.

Work is undoubtedly good, the essayist E. B. White once suggested. But he had evidently been spending a little too much time at the typewriter, because he went on to say: "I wish instead I were doing what my dog is doing at this moment, rolling in something ripe he has found on the beach in order to take on its smell.

"His is such an easy, simple way to increase one's stature and enlarge one's personality."

7

BENDING THE KNEE

Strategies for Subordinates

We admire our senior executives in the way that a black dog sticks faithfully by its beggar . . . even though he may be maniacal, lice-ridden, or schizophrenic. We blind ourselves to it, but our nose tells us what the world thinks of them. We have observed more closely than others the self-abasement and daily cruelty of which they are capable. And yet still we walk to heel.

—Robert Chalmers, *Who's Who in Hell*

We know by now how thoroughly a domineering CEO can hold his company's board of directors in thrall. The puzzling thing is that this thralldom casts its spell even over some of the most powerful individuals on Earth. At Hollinger International, for instance, the board treated Canadian newspaper tycoon Conrad S. Black with evident awe. In 2004, as part of a messy corporate divorce, Hollinger filed a report with the Securities and Exchange Commission charging that Black and other executives "willfully and deliberately looted" the company. Black replied that the report to the SEC was full of "exaggerated claims laced with outright lies." But no one disputes Hollinger's own admission that the members of its board performed their oversight duties with "lassitude," "acquiescence," and "passivity."

At one point, board members agreed to sell two of Hollinger's community newspapers to a private company owned by Black for $1 apiece, little more than an ordinary reader would pay to buy a single copy at the local newsstand. No one bothered to question the wisdom of the deal. Black's company quickly resold the two newspapers for $730,000. Another Hollinger paper disappeared into Black's pocket for $1 even though the company had a competing offer for $1.25 million.

At the same time Black was paying his wife, Barbara Amiel, a salary of $1.1 million as Hollinger's "vice president, editorial." Depending on which side you believe, Amiel's duties included either "participating in significant hiring decisions" and "providing editorial insights" or "ordinary activities such as reading the newspaper, having lunch, and chatting with her husband about current events." She was also a columnist at Hollinger's British flagship, the *Daily Telegraph,* where Black complained that she was underpaid at an annual compensation averaging $247,000.

When Black was writing a biography of Franklin D. Roosevelt, Hollinger generously helped out by purchasing Rooseveltiana, including personal papers, for a total of $9 million. Black did not actually get around to asking for the board's approval for one $8 million FDR purchase until two years after the fact, and then only as an item on the agenda of a seventy-five-minute teleconference. Even so, the board generously rubber-stamped the deal, based on the seller's assertion that the documents were worth at least as much. (The seller did not mention that he had paid $3.5 million for them less than a year previously.) The bulk of the collection apparently never crossed Hollinger's doorstep but served instead to ornament Black's private residences. After the final breach with Black, the company managed to get back the Roosevelt papers and sell them, for a $5.6 million loss. But a source "close to the board" told the *New York Times,* "It was not easy to say, 'You have to give the books back or sell them.' "

The directors who were apparently too cowed to make this simple demand included such eminent figures as former U.S. secretary of state Henry A. Kissinger, longtime cold warrior Richard N. "Prince of Darkness" Perle, and former Sotheby's CEO (and convicted felon) Alfred Taubman. Indeed, these titans appear to have done nothing to control Black until outraged institutional shareholders filed lawsuits charging Black and other executives with multiple counts of egregious self-dealing.

So what actually went on in the boardroom at Hollinger? The natural tendency is to focus on the alpha, and especially an alpha such as

Black. He is a bombastic character described in the report to the SEC as "infinitely skilled in reinforcing the aura" of his personal dominance. At the most basic level, he had established Hollinger as a public company under his absolute control, with a dual voting structure that gave him roughly 70 percent of the voting power even though his actual ownership share dwindled to just 19 percent. The company alleged in its SEC report that Black and other Hollinger executives further ensured that they got their way by feeding the board misrepresentations, omissions, and outright lies.

Black could also charm his directors and dazzle them with the glamorous social world where he and Amiel were bright lights. If the need arose, finally, he could scowl and bully into submission anyone who challenged him. When he met with the special committee belatedly appointed by the board to investigate his management practices, Black began by glowering at his inquisitors and threatening to sue them for libel, and to do so in Canada, where the libel laws were more to his liking. Richard Breeden, the former SEC chairman who was leading the investigation, later testified, "Mr. Black begins many conversations by threatening everybody."

But dominance is easy to understand. What about submission? How did such formidable figures as Kissinger or Perle submit so totally to such a willful boss, even to the point of risking personal liability for negligence? Corporate directors have traditionally been utter conformists: "Directors all have this ability to walk into a room and in ten seconds size up the culture and adapt to it. That's the selection criterion," says Nell Minow, founder and editor of the Corporate Library, an investment research firm. For the duration of a board meeting, they set aside their own dominance and become little more than mirrors reflecting the might of the person who brought them there. It is as if the captain has invited them aboard his ship, and it would be impolite, if not a flogging offense, to question his seamanship. This has been true at many companies even when the captain was steering a course clearly dictated by self-interest, with the shareholders drawn along haphazardly, if at all, like flotsam bobbing in his wake.

Much of the post-Enron debate over corporate governance has focused, quite reasonably, on reforming this traditional relationship by tinkering with the ways boards of directors are structured. Unfortunately, this discussion has seldom paused to consider how humans are structured. We don't necessarily subordinate ourselves to our alphas quite so abjectly as do other social animals. We are also more flexible, with the ability to move from group to group, and from the role of alpha in one context to beta in another. But we are primates nonetheless, and it is the nature of social dominance to distract us, despite our noblest intentions and best-laid plans, from issues of genuine substance.

When Henry Kissinger called power the great aphrodisiac, he was surely too much of a gentleman to be making a smug reference to his own knee-weakening effect on women. It seems far more likely that he was describing the intoxicating effect of other people's power on him. Or as the eminent columnist Barbara Amiel once wrote: "Power is sexy, not simply in its own right, but because it inspires self-confidence in its owner and a shiver of subservience on the part of those who approach it."

THE SHIVER OF SUBSERVIENCE

You do not need to look at the top of the corporate world to see how dominance affects the behavior of subordinates. Any human group or organization will do. Much as alphas tend to stand in an imposing way, or use a certain peremptory tone to address those around them, their underlings instinctively do the opposite. We acknowledge our subordinate status with posture, vocal tone, and facial expression, as well as in words. Most of the time we are not conscious of the submissive signals we are busily sending, and we may even deny any such behavior when someone points it out. Treating the boss differently from other people is "sucking up," and it's something only other people do. Little people, even. But researchers suggest that, to one degree or another, we all do it, as an automatic, unconscious response to the approach behavior of the alpha.

Let's start with an extreme example from the animal world. Chimpanzees sometimes practice submission so unabashedly that it's hard not to laugh, and also hard not to notice the embarrassing intimations of our own obsequious behaviors. Primatologist Frans de Waal describes one such underling pant-grunting, a sort of rapid, Kissingeresque, *oh-oh-oh* murmur of subservience, while at the same time assuming "a position whereby he looks up at the individual he is greeting. In most cases, he makes a series of deep bows which are repeated so quickly one after the other that this action is known as bobbing. Sometimes the 'greeters' bring objects with them (a leaf, a stick), stretch out a hand to their superior or kiss his feet, neck, or chest. The dominant chimpanzee reacts to this 'greeting' by stretching himself up to a greater height and making his hair stand on end. The result is a marked contrast between the two apes, even if they are in reality the same size. The one almost grovels in the dust, the other regally receives the 'greeting.' "

Among humans, this sort of thing sounds better suited to peasants kowtowing before feudal overlords than to the average Fortune 500 company, or even to the average Third World sweatshop. But humans merely send similar signals of submission by subtler means. People entering the presence of a powerful individual often stoop slightly and lower their heads. They glance down at the floor more often than one might expect based on the quality of the carpet alone. They may not bob, but they find frequent opportunity to nod. They don't grovel, but they tend to keep their hands at their sides and avoid expansive gestures other than a handshake.

When they come into the room late and the meeting has already started, they often perform an exaggerated mime of appeasement, stooping to appear smaller, tiptoeing, elbows at their sides, lips pressed against teeth in a smile of apology and chagrin. When the boss shows up late, by contrast, he is likely to arrive in a flurry, stride to the place of honor, spread himself out, and loudly commence business.

Consultant Bernie DeKoven has often watched bosses unconsciously make themselves larger while imposing a constricted posture on subordinates: The boss waits till the meeting has begun, then stands up to address the group and begins by taking off his jacket. Nobody else can get comfortable in the same fashion without also standing up and making a show of it, a small but potentially fatal act of insurrection. (Through his Institute for Better Meetings, DeKoven tried for more than fifteen years to make meetings "collaborative and fun." But he finally concluded that "meetings are ceremonies to reinforce the hierarchy, to remind people who's boss, and to praise or chastise anyone who isn't." If people would just admit it, he says, meetings could be shorter and cheaper.)

The facial language of subordinates also conveys deference. Subordinates are avid, if unconscious, practitioners of impression management. So they tend, for instance, to press their lips together more often, bite them, or suck on them, all unintended clues that they are trying to avoid displaying their true feelings. They smile more, to indicate cooperation and agreement. This doesn't necessarily mean that they are happy. (To paraphrase an old rubric about men and women, subordinates smile when they want to please; bosses smile when they are pleased.) They may well feel anger, but they are generally careful not to display it. They are quick, on the other hand, to show embarrassment or shame.

We like to think of these expressions as individual and entirely under our personal control. But they are in fact biologically scripted, in both muscle movements and social function. A display of embarrassment, for instance, almost invariably follows the same fixed pattern: First, look away. Second, try to suppress a smile. Third, break out in a smile despite yourself. Fourth, try to suppress it again. Fifth, drop head forward. Sixth, touch the face. Shame is simpler, a matter of dropping the gaze and lowering the head. These displays, psychologist Dacher Keltner notes, "resemble the appeasement displays of other species, which also involve gaze aversion, facial actions similar to smiling, head movements down, reduced physical size, and even self-touching or grooming."

Our response to these displays is also biological: Embarrassment elicits automatic sympathy and makes it less likely that the individual will suffer punishment. Think, for instance, about the mortified expression on a child's face after he knocks over a drink in a restaurant and how this can make you relent or even smile when you mean to be stern. Embarrassment, says Keltner, causes onlookers to laugh and "make light of the situation," thereby reducing the social distance produced by the mishap.

In fact, it's common for the embarrassment pattern to involve two more steps, in addition to those listed above: seventh, everybody laughs together, and eighth, they lean closer, as if drawn in by magnetism, and they touch. A show of embarrassment or shame demonstrates that the culprit recognizes social norms and wants to adhere to them.

And it works. People who commit social gaffes, children who misbehave, and even politicians who get caught in a lie or other transgression "are all liked more and punished less when they display embarrassment and other submissive behavior." In a mock trial staged by Keltner, a defendant convicted of selling drugs who showed embarrassment and shame got a shorter sentence and was put up for parole sooner than defendants who showed neutral behavior or contempt.

Words provide an even richer means of displaying deference. Insecure subordinates go out of their way to avoid strong statements or anything remotely controversial, because of the constant fear that conflict will undermine their relationship with the boss. (At Hollinger, according to the company's report to the SEC, "every Board member at least implicitly understood that Black would remove anyone who offered serious resistance to his dictates. That just wasn't done in *his* Company.") They soften their language with hesitations and with hedging phrases like "sort of" or "I guess." Declarative sentences pitch upward uncertainly toward the end, or they suddenly become questions with taglines like "don't you think?" The self-deprecating preamble is commonplace: "I don't know if this is going to be what you want, but . . ."

THE DEFERENTIAL CHALLENGE

Clearly, subordinate status is fraught with hazards. At one company, for instance, the boss removed shoes and socks at a meeting with two vice presidents and asked them to smell his feet. "He insisted that they smelled like strawberry shortcake," one vice president reported afterward. (Sadly, he did not reveal if he and his colleague actually bent down to sniff.)

A difficult boss can drive subordinates to ill-considered remedies. For instance, a California man working in behavioral psychology at a state hospital became familiar with B. F. Skinner's research. Skinner, whose narrow, mechanistic view of social life has fallen into disrepute, saw behavior largely in terms of conditioned responses, which proper social engineering and the right stimuli could encourage or suppress. Among other ideas, he proposed what he called "success of approximation": If you reward a dog each time he comes a little closer to the goal you have in mind, then sooner or later he'll actually get there. So the worker decided that he could use Skinner's technique to train his boss, like a bad dog, to treat him decently.

"And the way I did it is, I'd go in to work that morning," said the man, who proudly recounted his experience on National Public Radio's *Talk of the Nation,* "and I could see that I was going to end up being with group number three. It's not what I wanted to do today, but . . . I'd go to her first and say, 'Do you want me with group number three today?' and she'd look at my face and she'd say, 'Yes.' And [if] there was a mess on the floor over here, I'd go say, 'You want me to mop up that mess?' and she'd look at my face and she'd say, 'Yes.' And after doing this several hundred times, she got real good at saying 'yes' when she saw my face." He claims that he could ask for the afternoon off, and she would acquiesce. "We became buddies, and I really didn't need to put a lot of energy into getting her to say 'yes' anymore because we had a good relationship."

So there it is, easy as pie: If you are willing to clean up the worst messes left by the least appealing patients in an insane asylum, maybe your boss will treat you decently, too. Some oppressed employees, like some owners of bad(-ish) dogs, might well wonder who's training whom. Surely there must be a better way? Surely there are ways to pay respect without sniffing anyone's toes or flopping onto one's back in full wallow? There must be ways to be a subordinate and not end up like the "rather loyal and servile lieutenant" at one California high-tech company "whose Rorschach responses included poodles, an earthworm, and dissected testicles."

Spinelessness isn't in truth mandatory for subordinates. It is merely an innate tendency that many bosses cultivate, wittingly or otherwise, by the force of their personalities. Many companies encourage it, too, with the trappings of hierarchy and with manipulative, top-down management practices. It takes a secure, confident leader to do the opposite, encouraging subordinates to challenge his ideas, go their own way, or even get him to change his mind. The subordinate taking up the challenge needs an even greater sense of security and confidence (that, or a lot of what is sometimes referred to as "fuck-you money" in the bank). But even when both sides know each other well and enjoy a rapport, the deep evolutionary history of the relationship makes it prudent for subordinates to show deference and submission beforehand, to mitigate any hint of defiance.

GM vice chairman Robert Lutz describes his technique for challenging social superiors as a young executive at department meetings, when "big boss X" would summarily veto one of his proposals as "a dumb idea . . . and all the others, nodding furiously, were throwing stern 'how-could-you-be-so-stupid?' glances in my direction." With suitably deferential body language, Lutz would speak up in a tone of great reasonableness: "Excuse me sir, but we knew that 'no' was your original position, and still we felt there were some salient facts you weren't aware of. Because we think that, when we lay out all the facts, you may not want to reject it."

Lutz fondly recalls the "awestruck horror" on "the faces of the assembled apparatchiks. How could this incredibly unimportant person speak across five levels of hierarchy and *directly challenge* the big boss?" The answer, according to Lutz: "Easily, if you're tenacious, reasonably polite, have a ferocious belief that you are right, are proposing the right thing for the company, and believe that your arguments will convince a reasonable person." On second thought, hold the ferocious belief. Lutz, who named his business memoir *Guts,* understandably prefers to play down the way he established his deference first, in the manner of Dickens' Oliver seeking another bowl of gruel: "Excuse me sir, but . . ."

Not only should you tone down the ferocious belief, but you will probably be safer if you contradict or criticize the boss only in the privacy of his office. In public, your behavior may be taken as an attempt to incite open revolt. In private, it is more likely to appear as a declaration of both loyalty and independence. Leni Miller, head of EASearch, an executive-assistant recruiting firm in San Francisco, recently placed an assistant with a CEO. "She'd been on the job for a week and a half, and one day this guy just started screaming at her. She stood up, took him into the office, sat him down, and closed the door. She didn't do it in front of anyone else, but she said, 'You must never, never, *never* talk to me like that again, because it's just not possible for us to work together like that.' " Her tone was both sweet and a little disappointed, and it was a shock to the CEO because no one had dared to confront him before. He has not talked to her that way since. "If he starts, she gives him this look, and he understands the look. But an assistant can't do that unless she is very good at her game and knows that she is providing a service the other person can't live without."

Publicly rebuking the boss may make you a hero to your coworkers. But it is often also a form of career suicide. At ADP, the data processing company, a journalist was hired several years ago to write speeches for senior executives. The ad director called a meeting to try out his speech on his unfortunate staff, with the journalist present. At the end of a characteristically inept performance, he paused, looked up, and said to the journalist, "Jim, I gotta tell ya, this is *shit*!" A stunned silence ensued.

Then the journalist replied, "You're right, Barry." A nice start, suggesting a willingness to get past his natural emotional response to this abusive comment and work toward a practical solution. But then Jim added: "I wrote it that way because I knew you were going to deliver it." A second horrified silence followed. Then Barry broke out in a big, fake guffaw, and all his flunkies obediently joined in. (Jim worked for the company again, but never for that executive.)

If we cannot always defer to the boss with our words, we usually manage to do it with the tone in which we speak them. The moment of submission may be almost too subtle to detect, a preemptive surrender to avoid conflict. Researchers at Kent State University study very low conversational frequencies. What they're listening to is a deep hum, below 500 hertz, vocal but not verbal, which runs like a foundation beneath speakers' words. In every conversation they have studied, the low-frequency tones of the two speakers quickly converge. Their voices fall into step with each other, like two people walking.

They don't converge on some comfortable middle ground. Instead, they converge at the vocal level of the more dominant individual. That is, at some point in the conversation, one person submits and switches to the "power signal" of the other. It's an act of deference or accommodation, to reduce the stress of conflict and get on to the substance of the conversation.

The two researchers, Stanford Gregory and Stephen Webster, theorize that our vocal undertones provide a means by which we routinely and unconsciously manage "dominance-deference relations." Gregory recalls talking to one of his graduate students at a party when the dean briefly joined them. Gregory unconsciously shifted to match the vocal frequency of the dean, who, on some subliminal level, presumably expected this ritualized nod to his place in the hierarchy. When the dean left, the graduate student declared, somewhat indiscreetly (but no doubt with the proper vocal tone), "You just did it." This nonverbal form of communicating status, says Gregory, may be why one person overhearing a coworker on the phone can tell by tonal qualities alone whether she is talking to a boss or a friend.

POOH-BAH PROTOCOL, AND THE JOYS OF DEFENESTRATION

Does it matter if you omit these gestures? Most bosses will say, "Of course not. That stuff means nothing to me." But here is a handy rule of thumb: Most bosses lie. Friendly greetings and small gestures of deference are essential to the smooth functioning of almost any relationship, even between equals. They have become ordinary good manners because they signal benign intentions. They help us overcome our natural tendency to regard every new face as a possible threat. We trade small talk for a few minutes, or even for most of a conversation, because it helps us beat back the demon of negativity and get our shoulders down so we can proceed to the real business at hand.

Given the inherent inequality of the relationships in a workplace hierarchy, deference is even more essential on the job. Somewhere in the back of her mind, the boss is always wondering: *Is this person still my loyal lapdog? Or is today the day he wees on my carpet? Is this still my de-*

pendable colleague? Or is she about to abandon me for that job on the Henson account? And the employee in turn is thinking: *Will I get the raise she promised? Or is she about to ship my job off to Bujumbura?*

Prudent subordinates find ways to assuage the alpha. We go out of our way to include the boss in the conversation. We laugh at his jokes. We clear our throats outside his door, seeking an invitation to enter, much as chimps or baboons grunt when they want permission to approach. (Here is an exercise for executive assistants: Over the course of a week, keep track of how often the throat-clearing occurs in subordinates hoping to enter the offices of their superiors, and how often superiors bother to clear their throats when dropping in on subordinates. Bets, anyone?) At one investment firm, subordinates stand outside the CEO's office and air-knock to get his attention, rather than brazenly rapping on the glass door. These apparently meaningless acts of deference encourage workplace harmony.

Instead of treating this deferential behavior as spinelessness or sucking up, it may be more accurate to think of it as a natural element of any social relationship, and even as a healthy form of social manipulation. Smart subordinates routinely confer respect and status as a way to influence the behavior of the dominant individual, even down to the biochemical level.

In one intriguing experiment in the early 1980s, researchers from the UCLA School of Medicine separated dominant male vervet monkeys from their social groups, using a one-way mirror so the alpha could see his subordinates and make his customary threats and other displays. Dominant males normally have about twice as much serotonin in their blood as do their subordinates, and this tends to boost their confidence and also discourage destructive aggression. These are traits any sensible subordinate wants to cultivate in a boss. But because the one-way mirror prevented subordinates from seeing the alpha, they were unable to make appropriate displays of deference. This presumably infuriated the alphas, and over the sixteen days of the experiment, their serotonin levels dropped by 40 percent. In vervet monkeys, researchers Michael J. Raleigh and Michael T. McGuire later concluded, increased serotonin "facilitates a wide range of positive, prosocial behaviors.

Almost all of the conclusions about the impact of serotonin on social behavior derived from these studies can be generalized to humans."

Such studies suggest that our meaningless daily greetings are far more meaningful than we suspect. The boss may not care much about the verbal information when he walks into the office and says, "Good morning, how are you today?" But he cares about the tone of the reply, as a sign of whether all is well with the social order. Omitting greetings or the requisite note of deference can quickly lead to discord.

Watching his chimpanzees, Frans de Waal noticed that 60 percent of the serious fights between adult males occurred in periods when the individuals involved weren't greeting each other in the normal fashion. "These figures strengthen the assumption that 'greetings' have a reassuring effect," de Waal writes. "They probably serve as a kind of insurance for the dominant male that his position is safe. A show of respect in the form of 'greeting' on the part of the loser in the dominance process is the price he pays for a relaxed relationship with the winner."

What's the risk of failing to pay this price? Two cautionary tales suggest the potential hazards.

▪ In Connecticut a few years ago, the negotiator for state workers, Daniel Livingston, sat down with then-governor John G. Rowland and started out by calling him "John." Livingston later claimed that he was just trying to be friendly. Rowland tartly advised "Mr. Livingston" that he was being unprofessional. Negotiations broke off, and the next morning Rowland announced plans to lay off three thousand state workers. "It's not personal," said Rowland. "I've got a job to do." And Livingston said, "This is not personal with me." But all business is personal, and never more so than when people feel obliged to declare the opposite.

Might Livingstone have saved those three thousand jobs by addressing his counterpart as "Governor" or "Mr. Rowland"? Probably not. But his lack of deference apparently gave the imperious governor an excuse to bring the ax down sooner and with more enthusiasm. (Rowland later demonstrated just how personal business can be when he resigned in disgrace after admitting that deferential contractors had given him a hot tub, Cuban cigars, and expensive vacations—all good ways of manipulating his serotonin levels and making him favorably disposed when handing out lucrative state contracts.)

- Music producer Tommy Mottola also apparently struggled with the need to assume the subordinate position. He was, after all, the CEO of his division, commanding a small army of subordinates, trailing famous friends and beautiful women in his wake, with a $10 million annual overhead for travel, personal assistants, bodyguards, an armored car, and other perquisites. He liked to think of himself as the don, the *capo di tutti capi* of his world. Having started out as a promoter with his own Don Tommy Enterprises and risen to the top of Sony Music, Mottola couldn't bring himself to show deference to the head of Sony America, the avuncular and highly regarded former television news producer Sir Howard Stringer.

Something in his wise-guy personality made him flout authority, and Stringer naturally noticed. "It bothered Stringer," *New York Magazine* later reported, "that Mottola would never observe pooh-bah protocol by inviting him to the Grammys. A Sony employee remembers seeing Stringer at one Grammy party taking a group of people over to sit with Mottola and Mottola signaling his minions to close ranks because he didn't want Stringer sitting there. Mottola was known to refer to Stringer as 'the buffoon' in private. Stringer was also given mediocre seats at Sony-artist concerts, and backstage passes were usually out of the question.

" 'Tommy never wanted Howard at any event. He didn't want him near the talent,' says one former employee." Stringer used to phone his brother Rob, who was head of Sony Music UK, "and go, 'Why the fuck am I dealing with this guy? Why the fuck are we paying this guy? Tell me what this guy does. Tell me what this guy does. *Tell me what this guy does.*' "

Finally, falling profits in Mottola's division gave Stringer his excuse. He opened the drawn shades in Mottola's perpetually twilit thirty-second-floor office and booted him, metaphorically speaking, out the window. Then he issued the standard unctuous press release pronouncing Mottola "an icon."

An alpha chimp generally kowtows to no one. But human dominance is seldom so absolute. Our good standing in almost any hierarchy depends on submitting to the people above us, even as we are enjoying the deference of those below: Don Tommy not only refused to defer to Stringer, he also would not attend meetings with Sony bosses in Tokyo. By contrast, Stringer went out of his way to show respect (the Japanese phrase is *"atama ga sageru,"* meaning "lower one's head") to Sony executives. He invited Sony chairman Nobuyuki Idei to Davos and seated him at private dinner parties beside such New York luminaries as television interviewer Barbara Walters. When Idei resigned early in 2005, after a series of miscalculations, Stringer moved up to become one of the few foreigners ever to head a top Japanese company.

Any prudent leader learns to perform a delicate workplace minuet, displaying dominance behaviors one moment and submissive behaviors the next. Unlike chimpanzees, we don't have the luxury of sticking to one social group for most of our lives. A salesperson or a consultant may have to find her way into a half-dozen hierarchies at different companies over the course of a day. A senior executive going from meeting to meeting may need to do the same thing, and somehow establish a relationship with each group, all within a single company.

In addition to displaying deference ourselves, we can also profit from being alert to how other people display their subordinate status. It's a way to decode and work around the hidden agendas that shape the course of every workplace meeting.

ALL EYES ON THE ALPHA

When you sit for a while in a group of baboons or other monkeys, you will notice them look around at regular intervals to take in the social landscape. Their gaze often stops on the alpha. It's a way of reassuring themselves that nothing much has changed in their world: His Hairiness has not moved too far away, meaning the subordinate is still safely within his protective aura. Nor has he come disconcertingly close. This tendency of subordinates to watch the alpha can be a useful tool in a meeting with relative strangers. Often the alpha takes the head seat, introduces the major topics of discussion, determines how long each topic gets discussed, and sets the mood of the conversation—but not always. Some alphas slouch in a corner of the room and leave their subordinates to thrash out an issue. In those cases, the person who speaks *last* typically holds the real power, not the one who speaks first.

Fortunately, you can almost always identify a silent alpha long before that. People in meetings, like chimps in their troops, pay close attention to the head ape. They often direct their remarks to this covert alpha. They'll glance his way for hints of approval or disdain and proceed accordingly. When the alpha speaks, everybody else pays attention. On the other hand, when a subordinate speaks to the alpha, the alpha may well look away, or even exchange a comment with an aide, because he does not need to establish his deference.

We like to pretend that business decisions get made on the basis of facts, numbers, and rational choices. Everyone claims to want full and frank discussions, with even the janitor chipping in. But it seldom happens that way.

For instance, Silicon Valley companies often assemble software project teams including engineers, marketing experts, managers, and technical writers. The team leader is typically an engineer, who may not notice the way social status skews the discussion. Or if he does notice, he often lacks the management training or social skill to do anything about it.

So the dominance hierarchy "just sort of plays out, and the submissive sorts don't get to say anything, or their ideas get squashed," according to Jeff Johnson, a former software engineer at Hewlett-Packard and Sun Microsystems, who now runs a consulting firm called UI Wizards. Technical writers tend to be at the bottom of the hierarchy, and if one of them actually says something, people exchange puzzled looks. Then the conversation resumes without a ripple, often on a completely different topic. It's as if the wallboard has spoken, and everybody has tacitly agreed to pretend it didn't happen.

Johnson adds: "If a technical writer says something interesting, I'll stop and say, 'I think what she said is interesting.' And because I said it, and I'm an engineer and higher in the hierarchy, it will then be discussed."

When someone at the top of the food chain speaks, others will frequently rush to echo him, and the group will actively discuss the points he has just made. The urge to ape the alpha is powerful. Robert Lutz, for instance, describes how "big boss X" would grumpily read the memo about Lutz's "dumb idea" and declare, "Well, I guess that puts a different light on it. Is this what all of you think?" The same executives who had previously been throwing him "how-could-you-be-so-stupid" looks now responded with "an enthusiastic chorus of 'absolutely,' accompanied by much nodding."

It's a reminder of a disconcerting rule followed in almost every workplace: The status of the speaker matters more than the substance of what is said. Everyone pays attention to who gets "invited to play" with the big boys, says a former executive at Aetna, the insurance company. "In corporate parlance, if you're in the 'power curve' you are invited to closed-door meetings with the chairman and/or president. You are then 'at the table.' And you make sure everyone knows you were there."

Your opinions suddenly matter more. You matter more. "Another indicator of position is if the big guys solicit your views (as opposed to asking for facts) at meetings. And being thanked for your thoughts is like getting a big gold star. (Not a lot of thank-yous are uttered in big-guy meetings, I would note.) Conversely, if you have to ask for the floor, or don't get thanked for your pearls of wisdom, your perceived value drops. You are seen as bringing no 'value add' to the proceedings."

The urge to be "at the table" or "under the tent" often shows up as a powerful craving for physical proximity. At one company dominated for more than a century by the founding family, the chairman's son, a genial kid named Hovey, was working as an intern and being groomed to continue the dynasty. One day, he watched as the doors to headquarters flew open and his father came bounding down the stairs, closely followed by a vice president. Hovey watched them cross the courtyard in tight formation. "Boy," he said, "if Dad ever comes to a fast stop that guy's going to run right up his asshole!" It was an endearing remark, instantly repeated throughout headquarters. The embarrassed vice president thereafter was careful to follow at a more discreet interval, but never quite recouped his dignity. Hovey was last heard from playing in a band called Orange Splat.

BEING THE RIGHT-HAND MAN

In the animal world, subordinates pay close attention to the alpha partly to stay out of his way. It's the subordinate's job to step aside quickly when the alpha comes thundering down the trail. Smart subordinates become careful students of the boss's behavior so that they can recognize signs of danger even before a situation rises to the level of ritualized threat. Among animals, threat displays can vary considerably, not just from species to species, but also from one individual to the next. For instance, Frodo, the brutal alpha chimp in Gombe National Park's Kasakela troop, used to chew on his upper lip before committing mayhem, possibly an attempt to conceal his hostile feelings so he could land on his victim with greater surprise.

Noticing such cues is a matter of survival. Subordinates at EDS knew, for instance, that Ross Perot was about to blow when his left ear started going red. At one television production company, the top executive tended to rub his face with increasing intensity the madder he got. When newcomers missed this cue and continued pitching some inappropriate idea, old hands used to push back from the table and watch with an intellectual fascination otherwise reserved for police videos featuring horrific car accidents.

At one of the Wall Street companies where Sandy Weill was a rising alpha, "the back-office crew soon learned," according to biographer Monica Langley, "that they could interpret Sandy's mood and mind-set by watching his cigar. If he was putting the cigar in and out of his mouth and saying 'right-right-right,' he was listening. . . . If he was rolling the cigar back and forth between his lips and muttering 'kay, 'kay, 'kay,' they knew it was time to shut up or end the meeting."

The need to monitor and carefully interpret the alpha's gestures and expressions may explain the importance of being literally the "right-hand man." It turns out that we're better at picking up emotional cues about someone to the left side of our visual field, because these images fall on the right half of each retina and thus arrive more quickly in the right hemisphere of the brain, where we do much of our emotional processing.

Anthropologists Robin Dunbar and Julia Casperd first identified this phenomenon in male gelada baboons. They interpreted the tendency to keep a challenger on the left side of the visual field as an attempt to get a better sense of the rival's true intentions: "Is [the rival] bluffing when it makes heavy threats? Are the eyes that momentarily flicker away betraying a reluctance to press home an attack if pushed to the brink?" Likewise, when you sit with the boss (or your archrival) to your left, the greater sensitivity of the right hemisphere means you'll be quicker to pick up subtle nuances. Because you're on the right side of his visual field, conversely, he'll be a little less quick to notice what you're thinking.

Animals are also acutely attentive to their alphas because it's the subordinate's duty, often without being asked, to give up the shady spot he has been enjoying or the companion with whom he has been hoping to have sex. When the subordinate surrenders his place in this fashion, biologists call it a "supplant," and it is a common practice in the human workplace, too. Watch, for instance, the next time the CEO shows up in the fitness room looking to put in his half hour on the StairMaster. His favorite machine will miraculously become vacant, and the previous occupant will smile and say, "No, really, I'm just finishing," though he has not yet broken a sweat. Likewise, when the boss shows up at the door of an office where two colleagues are talking, one subordinate will typically smile, break off the conversation, and move away, supplanted. No one ever says, "We're busy now, could you come back later?"

Humans are more ambivalent about the supplant than other species, but the feeling that we ought to give way to the alpha remains strong. A road trip with the boss, for instance, presents a thicket of delicate questions. If he's bringing only a carry-on, does that mean you can't check your luggage? Do you have to have dinner with him every night, or can you go out with an old friend? If the boss doesn't drink, does that mean you shouldn't drink either? (There's always the minibar in your room. Then again, he may be the type to scrutinize the bill at checkout, note the six bottles of Bud, and loudly wonder if your *Hot and Naughty Nurses* rental really earned its XXX rating.) One business consultant, upgraded to first class, fretted about whether he should offer his seat to the boss, who was still back in coach. He finally made the offer, and his boss declined. Both of them had done the right thing. One of the functions of deference in a subordinate is to acknowledge the privileges of rank and enable the alpha to feel superior about forgoing them.

Failing to do so risks introducing an unspoken tension into the relationship. Not long ago, the CEO at a California computer security company checked into his New York hotel and got the same sort of luxurious room he had stayed in the week before, and again two or three weeks before that. But the head of investor relations, traveling with him, got lucky. Not only did the hotel clerk upgrade him, she put him in the presidential suite. It apparently didn't occur to Mr. Lucky to turn around and say, "No, you take it" to his boss. Maybe that would have seemed too obsequious, especially in front of a pretty young clerk. And it would have been brutish for the CEO to reach over and say, "Actually, I'll take that room." But the violation of the hierarchy of perquisites was duly noted. "You know, when it comes to being CEO, I'm pretty humble," the CEO remarked to his assistant, a few weeks later. "But . . ." Next time he went to New York, the assistant found him a new hotel with a keener eye for the social order. Later, for reasons that were no doubt wholly unrelated, the company found a new head of investor relations.

On the other hand, traveling with the boss can also offer the subordinate a window (or even an open door) into upper management. Staying close is a way to find out who's getting power, who's losing it, and how that's liable to affect everybody else. At one Fortune 500 company, for instance, a middle manager made it a point to catch rides whenever possible in the helicopters, planes, and limousines used by upper management. At one period in her career, she had to make a weekly trip to a company facility in a nearby state. It was a four-hour drive each way. Various vice presidents frequently made the same trip with far greater speed and comfort in one of the company planes, and the middle manager always tried to catch a ride with them. "You just have to know their assistants," she says. "The vice presidents have a better chance of getting the plane if they can show it's going to be full. So if you're nice to their assistants, they call you."

The middle manager's peers tended to be more inhibited about crossing the lines of hierarchy. "I told other people at my level, but they always said, 'Oh, it's not my place.' And I thought, 'No, it's your *job.* You're supposed to know what's going on with the company. This is one way you find out.' " One time, on a company helicopter, the middle manager introduced herself to her fellow passengers, who were all vice presidents. "Well, I'm not a vice president," she joked. It was a mildly abashed acknowledgment of having transgressed the social order.

The smoothest of the group, noting her relative youth, smiled and replied, "It's early yet." The two of them fell into conversation, and over the course of the flight she recognized that "he looked like a leader, he smelled like a leader." She also learned a tidbit the other vice presidents could not hear over the rotor noise: He was about to go off to a meeting attended by members of the company's board of directors. Back at headquarters the next day, "I told the vice president I was working for that this guy was going far, and he said, 'How do you know?' So I told him about the meeting with directors, and he said, 'He can't be.' " Soon after, Smells-like-a-Leader became the company's CEO.

IMITATION AND FLATTERY

Staying close and learning the boss's behavior makes it easier not just to interpret him, in all his alpha glory, but also to flatter and even imitate him. Despite what everybody says about hating flattery, it is an essential tool for subordinates.

Other social primates do their bonding mainly by grooming one another, and the choice of who gets groomed and who does the grooming is intensely political. Among vervet monkeys, for instance, a subordinate seeking to establish a relationship with a high-ranking female will often groom her ten times for every one time she gets groomed back. ("You are feeling deeply relaxed . . . your serotonin levels are peaking . . . you want to give me that banana . . .") If the subordinate is particularly shrewd or Machiavellian, she may also note hints of future power and go out of her way to groom a rising star. In one case, biologists in Kenya noticed that a relatively low-ranking female vervet monkey named Marcos got a disproportionate share of the grooming. Other females seemed to recognize that she was destined for power on account of her guile and her numerous family connections. And, in fact, over ten years, she gradually rose to the top rank. "The animals behaved as if they were hedging their bets," the researchers concluded, "forming bonds with those who had power at the moment but at the same time retaining bonds with those who might be powerful in the future."

People don't physically groom one another in the workplace, apart from the occasional back rub or the assistant giving her boss a once-over before he heads to an important meeting. We do our grooming with words. Humans evolved language, according to one leading theory, as a direct substitute for physical grooming behavior, and the grooming connection still shows itself in the words we use for flattery: *to curry favor, to butter up, to soft-soap, to suck up, to kiss ass.* The word *flattery* itself derives from a French word meaning "to stroke or caress."

In the workplace, we generally prefer both to give and get our stroking with words rather than with hands or lips. (This seems to be true even in contemporary Hollywood. One of the most prominent producers there put it succinctly, "I want the respect, not the blow jobs.") Human grooming also seems to be most effective, oddly, when it is least direct. Weak bosses surround themselves with yes-men who are always there to say, "Oh, you are the best. I can't believe how you did that." But most of us have evolved our Machiavellian primate heritage to the point that we mistrust those who flatter us to our faces.

So the shrewd subordinate places his flattery where the boss is likely to see it. James Truman, then editorial director of Condé Nast Publications, did it, for instance, when the *New York Times* called to talk about owner S. I. Newhouse: "Si has a real eye for the weakness of something," Truman remarked. "He won't say much, but just by standing there and going from one foot to the other while he looks at something, you know it needs work. His gift is knowing what makes people want to read magazines." Truman's gift, on the other hand, was in making inarticulate shifting from foot to foot sound like genius. If the *New York Times* doesn't happen to call, the shrewd subordinate will make his flattering remarks to a well-situated third party, confident that they will promptly be passed along to the boss.

Stanley Herz, who runs an executive recruiting company in New

York, once had an employee who complimented him endlessly, and did so in the highly effective context of client meetings, where it seemed like a natural part of the sales pitch: "This is Stanley Herz, and you need to know that he is the brains behind this business." Herz admits that he ate it up because the flatterer played to his need to see himself as an entrepreneurial whiz. To the chagrin of his other employees, Herz gave the flatterer a better job, a bigger bonus, fatter clients, and the freedom to take them to the best restaurants.

The flattery was so successful, in fact, that it eventually backfired. The flatterer was always an effective employee, says Herz. "I don't think anyone can fall for that kind of flattery if the person's not effective." But he became so expensive that he was no longer profitable to the company. When a recession forced Herz to put the brains of the business back to work, the flatterer was the first person he fired.

STRENGTH IN NUMBERS

One final element of subordinate behavior deserves consideration. When deference, flattery, and other forms of submissive display are not sufficient to make a boss do what they want, employees band together to do it with strength in numbers. People in meetings often wait to see if anyone will have the nerve to say what they are all thinking. Then, if it happens, they all pile on. It's almost like swarming behavior, with everyone depending on strength in numbers to keep them from being singled out individually for retaliation. Chimpanzees do the exact same thing. Thuggish behavior by the alpha provokes a female into an outraged "Waaa!" of protest, and a chorus of "Waaa!" promptly rises from her allies until the entire troop is caught up in a cacophony of indignation, forcing the alpha to back down.

Like chimpanzees, humans also routinely sort themselves into informal coalitions and may ultimately marshal their collective power to dethrone the alpha. At one school, for instance, a successful football coach regularly indulged in vicious personal attacks on his athletes. He also made a point of sowing animosity between individuals and groups. No one wanted to jeopardize his place on the team by standing up to this tyrant. Because they were constantly angry at one another, they also could not do it together. Finally, parents comparing notes in the parking lot intervened, and an informal coalition produced enough e-mailed instances of abuse to get the coach ousted.

At another institution, a boss regularly "pitched temper tantrums" at his subordinates. There didn't seem to be much the subordinates could do except commiserate with one another. But one day the boss actually threw his briefcase at an employee, and the ensuing indignant "Waaa!" from subordinates was loud enough to get back to the CEO. The creep did not get fired, but he suffered a 25 percent cut at bonus time.

Much of the history of labor-management relations follows this pattern of subordinates forming coalitions to level the power imbalance. We do it, for instance, with the help of labor unions, class-action lawsuits, fair-employment laws, sexual harassment laws, humor, and gossip. These leveling devices may seem like the product of human nature at its most progressive. But a couple of quick lessons from other primates can teach us a lot about coalition behavior:

- *Pay attention to how people connect.* When primatologists want to know which chimps are working together, they find out by recording who sits close together, who grooms whom and how often, how they share food, and who comes to defend another chimp in a conflict. You can make the same deductions at any meeting by paying attention to seating order, speaking order, and exchanges of facial expression. Political allies may even dress alike.

Desmond Morris has argued that allies also unconsciously echo one another's body language and that this "postural" echo can help define the factions in a meeting: "If three of the group are disputing with the other four, the members of each sub-group will tend to match up their body postures and movements by keeping distinct from the other subgroup. On occasion, it is even possible to predict that one of them is changing sides before he has declared his change of heart verbally, because his body will start to blend with the postures of the opposing 'team.' A mediator, trying to control such a group, may take up an intermediate body posture as if to say 'I am neutral,' folding his arms like one side and crossing his legs like the other."

- *Seek cues to the strength of a challenge.* Studying macaque monkeys,

Frans de Waal noticed that confident individuals concentrated their dominance challenges entirely on their opponent, as if there were no one else in sight. A less confident individual typically directed his efforts not just at the opponent, but also at potential allies among the onlookers. Such individuals called attention to their challenge with "exaggerated jerky turns of the head." A weak or insecure human challenger will typically engage in the same sort of "appeal aggression," confronting the alpha then nervously glancing around the room for allies.

If you are a subordinate, you should think twice about joining such an ill-prepared challenge. If you are the alpha, you should know better than to swat down the challenge. You will look stronger if you simply let it die on its own. Meanwhile, you can sort out the potential rebels from your loyal troops just by the way they respond to a challenge with a mute lifting of the eyebrows or a quick grimace of dismay.

> When a subordinate monkey wants to acknowledge and appease a superior, he or she may bend over, present hindquarters, and allow the dominant individual to mount and thrust for a moment or two. It isn't sex. It's just a polite way of saying, "You're the boss and I know it." Modern companies generally frown on this particular form of submissive display. But subordinates should never forget that, at many companies, the equivalent in words or deferential gestures is still often mandatory.

8

CHATTER IN THE MONKEY HOUSE

Gossip and the Beastly Secret of "Oh, My God, Tell Me More"

Live in such a way that you would not be ashamed to sell your parrot to the town gossip.

—WILL ROGERS

"Gossip Poisons Business," a headline in *Workforce* magazine declared not long ago. "HR Can Stop It." This is, of course, wonderful news. It suggests that the geniuses in the human resources department can also make crocodiles write sonnets and hippos dance the gavotte.

Gossip is human nature. Stopping it would take a miracle, or rather it would take black magic. And any company would grind to a halt in its absence. Despite gossip's reputation as the "devil's mouthpiece" and the "evil tongue," the flow of tidbits and scuttlebutt about the people around us is in fact our lifeblood as social primates and as corporate animals. Losing it would leave us socially and spiritually bereft. Indeed, one anthropologist has characterized gossip as a device for maintaining "the unity, morals and values of social groups." But this sounds terribly high-minded, given that the topic is gossip. So let's digress for a moment to something a little tawdry.

When he was deputy obituaries editor at the *Daily Telegraph* in London, journalist David Jones took delight in the passing of Lady Stevens of Ludgate. She was, he wrote, a "pneumatic brunette" best known for having introduced Britain to toe-sucking in her book *Woman as Chameleon: Or How to Be an Ideal Woman.* The obituary quoted her advice on "soothing" a husband after a hard day at work: "Always kiss your husband's body starting from his toes. After kissing his toes and sucking them (hopefully he has washed them), proceed to kiss every inch of his legs. . . ."

This was venturing into risky territory. One of the main functions of gossip in any workplace is to communicate the unspoken rules. On Fleet Street, one such rule forbids "slagging off" rival publishers, or their dead chameleons, and Lady Stevens had been married to the chief executive at the *Daily Express.* Jones blithely went on to note that journalists everywhere were delighted by her advice that wives should become their husbands' prostitutes because of "the insights it gave into the domestic arrangements of a newspaper baron." Jones' publisher might well have resented this as an aspersion on his own marriage and even more so because it put him in the position of seeming to dance on the grave of a rival publisher's wife.

But the obituary proved particularly indelicate because, at that moment, the *Telegraph* was negotiating a deal with Lord Stevens to print the *Express* and acquire a 20 percent ownership. Jones spent the next ten days in supervisory hell, trying to keep his head attached to his body and being shunned on the company elevators.

One seldom hears the lament "If only I had gossiped a little more . . ." But gossip is an essential tool for survival in the workplace. The trick isn't to stop it, except to put down the occasional malicious rumor, but to use it to maximum advantage. We gossip to stay on the qui vive about what the people around us are up to. ("Jones, did you happen to notice those chaps from the *Daily Express* on the lift the other day?") We gossip particularly to keep track of those above us in the hierarchy. This may be why bosses are so opposed to gossip: It's about them. We gossip for status, to be the one with the smug little smile and the chance to share the skinny. We gossip to learn the norms and values of our group. ("Did you hear what Jones in obits just did?") We gossip to control other members of the group and to discipline those who violate the norms.

Despite its obvious benefits, the prejudice against gossip still runs deep. Not long ago, the head of Dale Carnegie Training, a workplace-consulting firm, suggested that the best thing executives could do to increase workplace productivity would be to "ban water-cooler conversations." Exactly how, short of cutting out workers' tongues or turning them into robots? Um, well . . . "If you think about it," he admitted, "organizational politics have been around forever." And habits that have been around forever are sometimes hard to break.

IT ISN'T FAIR

Because they lack our capacity for language, monkeys don't gossip. But they come as close to gossip as they possibly can without actually being able to shriek, "Oh, my God, tell me more." They do it by eavesdropping. They pay acute attention to the behavior of the animals around them and use this social knowledge to win friends and influence fellow primates. Among other things, they seem to pay special attention to whether the organization is treating them fairly.

In a paper in the journal *Nature,* under the provocative title "Monkeys Reject Unequal Pay," researchers from the Yerkes National Primate Research Center described what happened when they introduced brown capuchin monkeys to the ways of the cash economy. Brown capuchins are a highly social and cooperative species native to South America. The "cash" in the experiment was a granite pebble, which the monkeys learned to exchange with the researchers for either a piece of cucumber or a grape. For the experiment, the researchers put two monkeys in adjacent cages, so each could see what the monkey next door got for her money. (By coincidence, all the monkeys in the "unequal pay" study were female.) If both of them got cucumber, they ate it happily and all was well with the world. But if the first one got a grape and the second one got a cucumber, organizational politics suddenly reared its ugly head. "They respond negatively to previously acceptable rewards if a partner gets a better deal," the researchers wrote. In fact, the outraged second monkey often howled in protest and flung the perfectly edible piece of cucumber out of the cage.

The Yerkes researchers, Sarah Brosnan and Frans de Waal, weren't particularly interested in simian gossip. But their experiment revealed one of the important reasons we pay close attention to other individuals around us, whether with our own eyes or by word of mouth: to make sure we aren't getting a raw deal. The capuchins were apparently feeling the same sense of violation we experience when someone gets a promotion without having earned it, or cuts ahead of us in line at the supermarket, or breaches the etiquette of alternate feed on the highway. The researchers suggested that the universal human "sense of fairness" and the aversion to inequity may be much older than we think.

These social emotions are powerful enough to induce the monkeys to hurt themselves, throwing away food for which they had paid good money, rather than accept an unfair deal. Despite the myth of "rational economic man," we humans don't behave any more logically, driving miles out of our way because we think the local shop once treated us badly, or quitting a job we love because somebody else got promoted ahead of us.

"People often forgo an available reward because it is not what they expect or think is fair," says lead researcher Sarah Brosnan. "Such irrational behavior has baffled scientists and economists, who traditionally have argued all economic decisions are rational. Our findings in nonhuman primates indicate the emotional sense of fairness plays a key role in such decision-making."

But if we gossip for practical reasons such as monitoring fairness and avoiding danger, some researchers suggest that we also gossip for the more positive purpose of strengthening the social bond with family, friends, and coworkers.

BIG BRAINS FOR BETTER GOSSIP

University of Liverpool anthropologist Robin Dunbar argues that gossip is the main reason, after eating, that we open our mouths in the first place. In a study among the presumably intellectual denizens of a university dining hall, he found that the conversationalists paid scant attention to ideas. Instead, they spent 70 percent of the time talking about one another. No other topic took up more than 10 percent of the conversation, and most, including "all the topics you might consider to be of great moment in our intellectual lives, namely politics, religion, ethics, culture and work" rated only 2 or 3 percent.

Dunbar suggests that it was roughly the same around Pleistocene campfires, because understanding social relationships has always been essential for survival. He argues, in fact, that we developed large brains, and the unique power of speech, essentially for the vital business of managing social intelligence. "In a nutshell," he writes, "I am suggesting that language evolved to allow us to gossip."

The human brain is about nine times larger than biologists would predict for a mammal our size. It's also our most expensive organ, consuming 20 percent of our daily energy budget. (Surprise: Your genitals are, comparatively speaking, a cheap date. Or they would be, if you didn't spend so much of your brainpower thinking about them.) Biologists figure brain size increased in different primate species only when it offered them a significant evolutionary advantage, to compensate for the extraordinary cost. A bigger brain might have helped, for instance, in mastering a larger home range, or obtaining different types of food.

Dunbar arrived at the idea that gossip was the critical advantage for humans by charting the brain sizes of different primate species, from dwarf lemurs to macaques. Then he correlated brain size on a chart with the size of the groups in which each species customarily lives. He found that the brain, and particularly the neocortex, the outer layer that does most of the higher-level thinking, got bigger in tandem with group size. The primates living in the biggest groups and with the biggest brains—baboons, macaques, chimpanzees, and humans—were the ones that spent most of their time on the ground, in relatively open savanna woodlands, or on the edge of the forest. This is rich habitat. But it's also a good place to get eaten by lions. So living in larger groups was essential to provide greater vigilance against predators. Larger groups also meant more social relationships. A bigger brain, according to Dunbar, became a necessity for managing those relationships.

Chimps keep track of their fellow chimps by spending a lot of time together grooming, and by paying attention to who grooms whom. They typically live in groups of about fifty-five animals, and grooming takes up about 20 percent of the day. Grooming helps individuals form social bonds, which can be vital when it comes time to share food or seek help in a fight.

More immediately, being groomed feels good. It releases a surge of endorphins and other endogenous opiates, the brain's built-in narcotics. We get the same blissed-out feeling in our own intimate relationships, when, as Dunbar puts it, language cannot adequately convey our emotions and we resort instead to rubbing, stroking, patting, and other forms of physical grooming (even perhaps the occasional toe suck): "As the endorphins triggered by these behaviors begin to flood the body, we experience a rising sense of warmth, a feeling of peace with the world, of well-being toward those with whom we share such experiences of intimacy. The effect is instantaneous and direct: The physical stimulation of touch tells us more about the inner feelings of the 'groomer,' and in a more direct way, than any words could possibly do."

GOSSIPING TO FEEL GOOD

The trouble with physical grooming as a form of bonding for humans is that we know too many people. And we generally don't want to be rubbed, stroked, or patted by them, especially not at work. By Dunbar's calculations, the human brain is geared to a group size of about 150 people. If we had to maintain our social relationships in such a large group with mutual grooming, it might eat up 45 percent of the day. We wouldn't have time to do important stuff like going to lunch. So instead, we keep track of one another and comfort one another with language, mainly by swapping gossip.

Language itself is surprisingly positive: One study of thirteen languages found "a universal human tendency to use evaluatively positive words more frequently than evaluatively negative words in communication." Another study found that we naturally tend to position the positive term first in standard word pairs—win or lose, more or less, happy or sad, us and them, profit and loss.

Gossip itself is actually negative only about 5 percent of the time, according to Dunbar. Far more often, it's just idle chatter: "Did you see that blouse she wore yesterday?" or "He's got a nice way with words." The content of gossip matters less, says Dunbar, than the "message of social commitment." When someone pauses to gossip, he or she is including you both in the one-on-one relationship and in the social group. The word itself implies as much. People who gossiped together were originally "god-sibs," like godparents, keeping each other close with the mutual sustenance of social news.

If Dunbar is right, then gossiping, like grooming, ought to make us feel good. Circumstantial evidence suggests that it does. Gossip and the sense of being included in a social group are surely among the reasons instant messaging, and to a lesser extent e-mail, is so addictively fascinating for teenagers. It's why people so often have cell phones to their ears, so they can engage in chitchat with friends, family, or fellow workers even while driving alone in the car or walking through a crowd of strangers on a city street.

AT&T shrewdly exploited the idea of idle conversation as something akin to physical grooming with its celebrated "Reach out and touch someone" advertising campaign. "Remember," said one ad, "no matter how far away your family, or family of friends, may be, you can always reach out and touch them." Unfortunately, no one has yet demonstrated that gossip, like grooming, actually boosts the production of opiates in the brain. There is likewise no evidence that such an effect can occur when the social exchange takes place through the relatively impersonal medium of a telephone or a computer.

We know, however, that the opposite occurs: Our brains respond to social exclusion much as they respond to physical pain, and this feeling of hurt occurs even when we're being left out by a computer. In one study at UCLA, a test subject played a virtual ball-tossing game via computer link with two other players. The other players were virtual, but the test subject didn't know that. They tossed him the ball seven times. Then they shut him out for the next forty-five throws, and he reacted as if he had been slapped.

In males and female test subjects alike, an MRI showed activity in the right ventral prefrontal cortex, the same part of the brain that gets activated by danger and other negative stimuli, and also in the anterior cingulate cortex (or ACC). The ACC is best known as a "neural alarm system" for detecting pain. But it also plays a role in monitoring social relationships.

In squirrel monkeys, for instance, it's part of the neural machinery involved when an infant produces a separation cry as a way to reestablish contact with the group. In humans, it plays a role in eliciting the distinctive twinge a mother feels when her child cries out. The UCLA researchers theorized that our neural systems for sensing pain have been co-opted in the course of evolution and put to work to promote social connectedness, so they detect when "something is wrong" in our social world. A social snub can thus be literally painful, like a stubbed toe.

A more recent experiment at the University of New South Wales didn't even bother to pretend that there was anything but a computer on the other side of the ball-toss game. But test subjects still experienced a sense of injury and ostracism. Or as the headline put it on their article in the *Journal of Experimental Social Psychology:* "How low can you go? Ostracism by a computer is sufficient to lower self-reported levels of belonging, control, self-esteem, and meaningful existence." The authors interpreted their results "as strong evidence for a very primitive and automatic adaptive sensitivity to even the slightest hint of social exclusion."

It is perhaps a leap to suggest that the feeling of being slighted during a computerized ball-toss game also occurs when we get left out of the office grapevine. Then again, why shouldn't it? In some ways, the content-free character of most chitchat makes it not all that different from a ball-toss game; it's more about doing something together than the particular thing being done. But gossip can also contain information vital to our survival on the job. And whether the topic is trivial or momentous, all of us at one time or another have experienced the visceral feeling of being left out of a conversation. So the potential ought to be much greater for gossip to produce feelings of being either part of the group or excluded from it.

The worst sort of gossip, an insurance executive admits, is "to hear about something important from another executive who is your peer, because that means he or she got to talk to the big guys and *you didn't.*

For power-curve players dying to be at the table and under the tent, being out of the loop is the kiss of death."

NINE STEPS TO BETTER INSIDE SKINNY

What are the practical implications of these ideas in our working lives?

▪ No matter how HR tries to stop it, people are always going to gossip. You have about as much chance of stopping them as you do of stopping your dog from sniffing other dogs' bottoms.

▪ If you're a manager and you engage in favoritism or other unfair practices, your subordinates are almost certainly going to find out. So don't do it, and be prepared to provide a credible explanation when you promote one person and not another.

▪ If you're hearing dangerous or demoralizing gossip, you should deal with it honestly: "We've all heard that the company is looking at whether it can do this work cheaper in Malaysia. What I know is that we still have a chance to show them that we can do it better, and at a competitive price, right here."

If you tell them that offshoring isn't even on the table and two months later ask them to start packing equipment for Kuala Lumpur, no one will ever again believe anything you say. Instead, they'll place even greater faith in the grapevine. (Except that the grapevine will have moved to Kuala Lumpur, too. And your job with it.)

▪ Conventional wisdom to the contrary, managers should use gossip as a tool for creating a sense of inclusion among their subordinates. The people in HR are right about one thing: This doesn't mean sharing your complaints with subordinates. It doesn't mean venting to them about the stupid, comic-book-reading slob of a production assistant who isn't pulling his weight and never has. In some jurisdictions, in fact, a company may have to pay substantial damages if the aforementioned slob successfully sues a manager for subjecting him, via gossip, to a hostile work environment.

Managers seldom get anywhere good with negative gossip. But again, most gossip isn't negative. You can help colleagues use their time more productively and also win their loyalty with a timely word: "I hear something big is coming down on the Disney account. Maybe you should hold off on that pitch till after the weekend." Be aware, on the other hand, that it's important to pass on the information to your subordinates more or less equally.

▪ Gossiping with people at the bottom of the ladder is not just a good way to get them invested in the group. You may also learn something, because people at the bottom often see more than people at the top. At one institution, the board of governors had a security guard listen in via headphones from his post in the hallway, so he could rush in if some geezer keeled over upon reading the chairman's compensation package. But he also heard everything, and sometimes shared it with people smart enough not to look down their noses at a security guard.

At a Fortune 500 company, a middle manager with a keen ear for gossip got a ride home one night in a company limo. She was alone, presumably a disadvantage for information gathering. So she sat up front and chatted with the driver. They talked about a young, attractive female project director named Nancy who was rocketing through the corporate ranks. Other project directors were muttering that she was incompetent, avoided decisions, changed her mind frequently, and communicated poorly—and she still got promoted.

The driver didn't explain outright that the project director was having an affair with a senior vice president. But it came out that the senior vice president's wife had hired a private detective, and when the senior vice president and Nancy, who was also married, traveled together on business, they took care to book separate flights arriving at different terminals. Nancy is now a senior vice president, too. For the middle manager, this gossip was not directly useful; she wasn't planning to blackmail anybody. But it gave her useful insight into corporate Kremlinology.

▪ Gossip like that can tell you where to invest your energy and with whom. At one point, for instance, Sun Microsystems set up a separate entity, a "skunkworks," to develop new software. The team leader was an engineer. But the company also brought in a managerial type, and the two inevitably clashed. The staff made bets on which one would force out the other. The smart money, knowing the nature of the corporate world, figured that the manager would prevail. "But another guy happened to know that the engineer played hockey with CEO Scott McNealy every week. And in the end, it was the manager who went." The smart money was wrong and, as is often the case, gossip got it right.

▪ For gossip to be any good, as either a form of bonding or a source of news, the relationship must be reciprocal and marked by a degree of trust. Trust, but of course with your eyes open: At one large company, a middle manager makes a point of cultivating colleagues as spies and allies. She tests them first to establish trust, by leaking a piece of useful information, something that won't be too damaging if it gets around. Then she waits to see if they heed her request to keep it confidential, and also "if they understand that reciprocity is expected."

Trust also means making your gossip as accurate as possible. Despite its reputation as a source of wild rumors, studies suggest that information on the company grapevine is typically 75 to 95 percent accurate. It wouldn't make much sense for people to invest energy in spreading false information, because it inevitably discredits the source. Even so, carefully check out rumors (especially e-mail rumors) before deciding whether to pass them on. If you hear something big is coming down on the Disney account, and you interpret this to mean that Minnie and Daisy are about to announce their gay interspecies marriage, you will soon wither on the company grapevine.

▪ Though top executives pretend to disdain gossip, they should also know how to use it as a competitive tool. A few years ago, for instance, an American insurer found itself pitted against one of the world's largest corporations in a bidding war to take over a European target. The insurance company figured it was being used as a stalking horse by investment bankers to drive up the bid by GigantaCorp. It decided to bid anyway.

But along with its bid, it shared a crucial piece of gossip with the target: GigantaCorp's standard tactic was to bid high, drive out the competition, then nickel-and-dime the target back down to earth with contingencies and sub-subclauses. The insurance executives also stipulated that they wanted to meet during the bidding process with their

European counterparts. This brief chance to chat (and gossip) face-to-face established a rapport, and the European company accepted the lower bid.

▪ Gossip can also serve to even things up on a personal level. It's one way we keep each other honest. Primatologist Christopher Boehm characterizes gossip as "a stealthy activity by which other people's moral dossiers are constantly reviewed . . . In spite of the secrecy, everyone knows that gossiping is constantly taking place; anyone can be a target. This knowledge serves as a deterrent."

At one company, for example, an executive got picked up on a Monday morning by the company helicopter and thus accidentally learned that the CFO was hitching a free ride every weekend to and from his vacation home at the beach. Nobody had to say it out loud. They just looked at each other, and the CFO knew she knew. On the other hand, people took very loud note when they found out about Richard A. Grasso's outlandish pay package as chairman of the New York Stock Exchange. "You see, he got the grape, and we got the cucumber," the *New York Times* commented, with the Yerkes research freshly in mind. "No. Let's try that again. He got a bunch of grapes, and a banana, and a hot fudge sundae with a cherry, and a Bentley. He got—let's spell it out—ONE HUNDRED THIRTY-NINE MILLION, FIVE HUNDRED THOUSAND GRAPES. The rest of us are sitting here, good little capuchins with our cucumbers. And our cucumbers are smaller this year. Cutbacks, you know. Tough market. Sorry."

It is probably smarter in the long run, though not nearly as cushy, to live as Will Rogers suggested, with the idea that your parrot has an open line to the town gossip.

In the movie industry some years ago, for instance, a snake of a Hollywood agent (the phrase may be unfair to snakes) was representing a young actor from a well-connected acting family. So when the agent bought a new answering machine, he passed on his old machine to the actor's mother. The answering machine turned out to be his parrot. The agent had stupidly left a tape in the machine of a conversation with a director about an upcoming project. The mother found herself listening in and soon heard her child being described as a dreadful actor and all wrong for the project. The agent was touting another client instead.

So the mother phoned up the agent, played back the relevant passage, and told him she would leak it to the Hollywood press if he did not immediately tear up her child's contract. The dreadful actor in question quickly found a new agent and is now one of the biggest and most highly praised names in popular entertainment. And the agent? He still looks like some kind of Hollywood reptile, only not as smart.

NOISES OFF

Paying attention to their fellow actors isn't the only way animals get ahead in the world. It's also about recognizing little signals offstage. One day in Botswana, I was watching a troop of baboons walking across a field, the rounded tops of their heads just cresting the high grass. They were walking straight toward a cheetah, which was advancing on them at a crouch. Then a bird, a hornbill, spotted the cheetah from its perch in a tree and gave an alarm call. The lead baboon looked up, figured out where the danger lay, and veered off on a safer course. His cover blown, the cheetah veered off, too. (A good thing, really; cheetahs are no match for baboons.) Predator and prey never actually saw one another. They simply knew how to read the changing landscape.

Office chitchat serves the same function. A contract worker at one company picked up, from something in a coworker's tone of voice, that the company was about to be sold. So the worker stayed up nights to deliver his project early. The result was that the manager who had commissioned the job paid for it (plus a bonus, because the bottom line hardly mattered anymore). The alternative might have involved getting eaten for lunch by some new manager eager to display his fierce cost-cutting instincts.

9

BANG BANG, KISS KISS

The Natural History of "I'm Sorry"

This is like the marriage of two porcupines. They will have to go about it very carefully.

—A SILICON VALLEY OBSERVER ON THE PUBLIC RECONCILIATION BETWEEN SUN AND MICROSOFT

One morning not long ago, an employee in the London law office of Dewey Ballantine sent out an e-mail offering a litter of puppies. A high-ranking partner then chipped in a waggish reply to the entire firm urging, "Don't let these puppies go to a Chinese restaurant!"

It was a dumb joke. Moreover, the firm had just gone through sensitivity training after having infuriated the Asian community the year before. The prior infraction involved a black-tie dinner, where laugh-a-minute lawyers had marked the closing of the firm's Hong Kong office with a song-and-dance: "You were the firm's folly and now we so solly to be cutting off your source of livelihood." It looked like a pattern of bias, and an apology failed to mollify outraged critics.

"Somebody made a mistake, and they've apologized," a cochairman at the firm lamented. "And we keep apologizing . . . I wish that there were some way that we could convince them very easily and quickly that this was truly aberrational with respect to our culture. But clearly that's not going to happen." It did not seem to occur to him that expressing a wish to do it easily, quickly, and in an "us-them" context might just worsen the problem.

Attempting to correct a social gaffe with a gesture or a

phrase is always a delicate business, and few transactions seem to be so quintessentially human. But the art of saying "I'm sorry" has deep biological roots, and they affect even our most nuanced apologies. Understanding those roots is important, not least to remind ourselves that people in any successful relationship need to hear the words "I'm sorry," and give voice to them, much as they need to eat and sleep.

It's especially important in a conflict-prone environment such as the workplace, where we often seem to apologize so ineptly, if at all. For instance:

- Cincinnati Reds star Pete Rose not only placed illegal bets on baseball games, including games where he was a player or manager, but also lied about it for fourteen years. When he finally came clean in 2003, his apology was tainted by self-interest: Rose wanted to lift his lifetime ban from the sport. He wanted to get elected to the Baseball Hall of Fame before the end of his eligibility, in December 2005, and he wanted to sell his self-pityingly titled book, *My Prison Without Bars.* In the book, for the record, this is the closest he came to an actual apology: "I'm sure that I'm supposed to act all sorry or sad or guilty now that I've accepted that I've done something wrong. But you see, I'm just not built that way. So let's leave it like this: I'm sorry it happened and I'm sorry for all the people, fans and family it hurt. Let's move on." It was something less than a Charlie Hustle, all-or-nothing headfirst bid for forgiveness.
- In 2004, after an independent commission found that "defective" editorial policies had allowed a BBC reporter to air "unfounded" allegations against the government, the British broadcaster's top two officials resigned, and the new acting chairman apologized "unreservedly." (According to the commission, the reporter lacked evidence for his claim that Prime Minister Tony Blair's staff had "sexed up" an intelligence dossier to mislead the British people into the Iraq war.)

This apology caused outrage among the twenty-seven thousand BBC employees, many of whom regarded the commission as a "whitewash" of Blair policies. They barely knew their acting chairman and felt he lacked the standing to apologize on their behalf. The spirit of that apology, which the BBC's own Sir David Attenborough called "groveling," nonetheless seemed to permeate the organization, with embarrassing effects. At one point, the BBC actually apologized because its most famously persistent television interviewer had been, of all things, too persistent in questioning the police chief in a notorious murder case.

At this point, it may well seem that we learned the art of apology from monkeys, and not very clever monkeys at that. But apologies are serious business. They are, or ought to be, the balm of our working lives. They have the potential to heal. They can suck hatred and fury out of the air. They can miraculously transform a customer's hostility into honey. They can salve a mother's fury over the needless death of her child and somehow turn her into a friend.

They can also have a dramatic effect on the bottom line:

▪ In 1987, the Veterans Affairs Medical Center in Lexington, Kentucky, established an "extreme honesty" policy for its staff. This has at times meant admitting medical errors, apologizing for them, and initiating a claim where the family itself had no reason to suspect that medical error contributed to a patient's death. Skeptics predicted that the policy would produce a liability nightmare. But legal costs at the hospital soon went from among the highest in the VA system to among the lowest. An honest apology apparently worked better than money to relieve the sense of suspicion and injury in the hearts of malpractice victims and their families.

▪ At Johns Hopkins Hospital in Baltimore in February 2001, an eighteen-month-old named Josie King was recovering from an accidental scalding. Shortly before the child's scheduled release, her mother, Sorrell, noticed that Josie had suddenly become dehydrated. The staff rebuffed two requests to intervene and, after a third request, mistakenly administered a drug that, together with the dehydration, killed the child.

The parents were bereft and enraged. Then a hematologist named George Dover, who was the director of the Johns Hopkins Children's Center, visited the Kings at their home. "We were involved with our lawyer then. We were going for it," says Sorrell King. "If George had said, 'We're not sure what happened,' we would have thrown him out. But he totally did the right thing, at least from our perspective. He said, 'I am so sorry. This happened on my watch, at my hospital. I will help you get to the bottom of it.' " The apology began a rocky relationship, with the two sides working together to investigate the causes of Josie's death. The Kings eventually settled their lawsuit. Then they donated a portion of the settlement to hire a pediatric safety officer and make other reforms in the unit where their daughter had died.

- In a case of serial sexual abuse by a Texas priest, the victims agreed in 1998 to forgo a jury award that had grown, with interest, to $175 million, settling instead for $23 million. There had, of course, never been any guarantee that the original judgment would have survived on appeal. But the breakthrough in negotiations came when the Catholic bishop of Dallas agreed to issue a personal apology and take steps to prevent future cases of abuse. "The pain will never go away," said one victim. "The dollar amount can't help that . . . the only thing that can is seeing the diocese change." The bishop's apology served as a public promise of that change, the outward sign of an inward transformation.

THE SACRAMENT OF RECONCILIATION

Primatologist Frans de Waal first discovered that there is a natural history of "I'm sorry" in the mid-1970s at the Arnhem Zoo in the Netherlands. His epiphany occurred immediately after a raucous confrontation (described in an earlier chapter) between Nikkie and Luit, two high-ranking males. As chimpanzee dominance contests go, this one was all bluster, all ritualized sound and fury and kicked-up dust. After a round of screaming and chasing, the two rivals ended up safely separated near the top of a dead oak tree. About ten minutes after the fight, Nikkie, the alpha male in the group, reached out his arm toward Luit, fingers extended, palm upward, in an offering of peace. De Waal snapped a photograph, and watched as the two apes descended to the fork of the tree, where they kissed and embraced, then climbed down to the ground and groomed each other.

Most researchers would have consigned this moment to oblivion under some dusty academic category like "postconflict interaction." Discussing animals in language that hinted of emotion was heresy back then. But de Waal decided to describe what had happened between Nikkie and Luit with the same word he would have used after a fight between his own brothers: reconciliation.

After that, de Waal began paying attention to how chimpanzees manage to live together before and after their frequent fights. He kept systematic records of their behavior in the aftermath of conflicts—the little apologetic gestures, the efforts to repair disturbed relationships—and he discovered that chimps are champions of reconciliation. Over the years, he has recorded more than forty-five hundred antagonistic incidents within his study groups—along with ten thousand grooming bouts and one thousand reconciliations. Chimpanzees might sometimes be "killer apes"—in fact, Nikkie ultimately killed Luit—but far more often deWaal's chimpanzees look like natural-born peacemakers.

This is *chimpanzee reconciliation lesson number one:* When they fight, chimps kiss and make up, and we should, too. It may sound simplistic to say so. But when he began to research the subject, de Waal found that science was "virtually ignorant of reconciliation behavior in private human relationships."

Social psychologists studying human behavior often did so in laboratory experiments, rather than among natural social groups, so they hardly ever saw reconciliation behavior. Family therapists routinely encouraged reconciliations, but they did so under supervision, in highly artificial circumstances. So they had no reason to think it was a natural behavior rooted in our biology. "There is no forgiveness in nature," the Italian playwright Ugo Betti once declared, and even romantic fiction denied the importance of reconciliation. A best-selling novel once popularized the absurd premise that "Love means never having to say you're sorry." Corporate executives likewise have built careers on the motto "Never apologize, never explain."

The truth, surprisingly, is that the natural world overflows with the spirit of apology. Since de Waal's epiphany with Nikkie and Luit, researchers have found forms of reconciliation in twenty-five primate species, and also in far more distantly related animals, from dolphins to hyenas. It happens not just among the members of a species, but even between separate species, as when the family dog comes creeping back in search of forgiveness after you've caught him snatching food off the kitchen counter. If you have the slightest doubt about how important a gesture of reconciliation can be, pay attention next time to the way the cringing and tension evaporate from the dog's body the moment you finally relent and pat him on the head.

Now keep that thought in mind next time a subordinate screws up. When you say something conciliatory—"Look, I'm sorry, I shouldn't have yelled at you like that"—you will see the subordinate's clenched body unwind with instant relief. Listen closely, and you can hear the ebb tide of cortisol retreating from the beachheads of the mind. You will hear it, that is, if your gesture of reconciliation is sincere.

This doesn't necessarily mean saying the substance of your complaint was wrong and the thieving dog or the screw-up subordinate was right. All it takes is a kind word or (careful now) a reassuring touch that says, "This relationship matters more than that. We'll figure out a way to move on from here."

The willingness to make such a conciliatory gesture is, in fact, often a mark of strength, not weakness. And if Nikkie and Luit could do it in the treetops, surely we can, too.

IRRECONCILABLE DIFFERENCES?

"Humans," de Waal writes, "make up in a hundred different ways: breaking tensions with a joke, gently touching the other's arm or hand, apologizing, sending flowers, making love, preparing the other's preferred meal . . ." Like chimps, we also kiss, "the conciliatory gesture par excellence." This may not be acceptable in the workplace, but de Waal notes, "Another point in common with the chimpanzee is the critical role of eye contact. Among apes it is a prerequisite for reconciliation. It is as if chimpanzees do not trust the other's intentions without a look into the eyes. In the same way, we do not consider a conflict settled with people who turn their eyes to the ceiling or the floor each time we look in their direction."

A researcher at one company recounts an incident, for instance, in which a manager ordered her into the top boss's office on a Friday afternoon and delivered a severe reprimand, with much loud talk and finger jabbing. The researcher refused to accept the reprimand. "I think you're completely off base," she told the manager. Much worse from the manager's point of view, the top boss pursed his lips and raised his eyebrows to show that he didn't buy his manager's accusations either.

On Monday, the researcher and the manager passed in the hallway, and she gave him her usual hello. He blew past with eyes fixed on something in the ceiling ductwork. He could not bring himself to resume normal eye contact, and having been falsely accused, the researcher felt she had no grounds for apologizing. After fifteen years on the job, she began sending out her resumé.

They might have avoided this unfortunate development with a little help from *chimpanzees reconciliation lesson number two:* It's sometimes possible for prideful antagonists to come together with the help of a "collective lie" so that neither party loses face. Proximity and the focus on a task unrelated to the dispute may then break the tension and lead naturally to a rapprochement.

In the aftermath of a fight, for instance, one of de Waal's chimps would frequently pretend to have discovered something in the grass "and hoot loudly, looking in all directions. A number of chimpanzees, including his adversary, would rush to the spot." The others quickly lost interest in this trumped-up discovery. But the two antagonists tended to stay together, focusing their attention on the object with a show of excited sniffing and hooting. "While doing so, their heads and shoulders would touch. After a few minutes the two would calm down and start grooming each other. The object, which I was never able to identify, would be forgotten."

The anecdote is a useful reminder that both sides, transgressor and victim alike, typically feel the powerful urge to reconcile. In human relationships, unfortunately, the wife or the female party to a dispute often gets stuck with concocting the collective lie. The male could of course also do it. He just doesn't. In the case of the female researcher, her awareness that the "emotion work" was being distributed unfairly by the manager who had reprimanded her merely added to her sense of umbrage, making any reconciliation more difficult.

Apes don't apologize all the time, either. Mountain gorillas make conciliatory gestures about 37 percent of the time after a conflict, captive chimpanzees do it about 40 percent of the time, and their close relatives the bonobos, those simian love puppies, reconcile almost half the time. Some people believe de Waal overstates the case for peacemaking.

One critic remarked acidly that he wasn't sure, at first, how these "cloying chimps . . . fit with the David Attenborough footage I saw the other night in which male chimps tore shrieking baby monkeys to pieces and then ate them, until I remembered they shared the bloody bits with their mates. Cooperation is beautiful."

But de Waal is talking about conciliatory behavior within the troop, an entirely separate issue from killer ape behavior toward a prey species, or even toward chimpanzees in rival troops. Any corporate manager should understand the value of fostering unity within a team, the better to tear the competition to bloody bits. Other critics suggest that de Waal may overstate the rate of reconciliation even within groups because he studies only captive animals. (Then again, captive chimps are probably a better model than wild ones for the behavior of the corporate animal.) "I think he is not fully aware of the importance of unreconciled aggressive interactions in the wild—or he chooses not to recognize them," says Richard Wrangham, whose book *Demonic Males* uses ape behavior to explore the origins of human violence.

When chimps quarrel in the wild, Wrangham says, they reconcile less than 13 percent of the time, only about a third as often as in captivity. This may be because the combatants can get away from each other in the wild, or because, for strategic reasons, they choose not to reconcile. The statistic may also be skewed by the difficulty of actually observing chimps in the wild after a fight.

But Wrangham promptly adds the conciliatory note that he "totally admires" de Waal for having demonstrated that primate life isn't exclusively about waging war, nor about making peace. "If you know chimps, you know they do both."

PORCUPINE LOVE

In April 2004, Steve Ballmer and Scott McNealy, the chief executives of Microsoft and Sun Microsystems, stood on a stage together and made peace. For more than a decade, McNealy had been denouncing Microsoft as "the beast from Redmond." He had nicknamed Ballmer and Microsoft chairman Bill Gates "Ballmer and Butt-Head" and attributed their refusal to meet him for a public debate to their inability to "even flirt with telling the truth." Ballmer in turn had characterized McNealy as "two standard deviations away from reality" and Sun as "just a very dumb company" staffed by "sub-50 IQ people."

The rivalry went well beyond name-calling. Microsoft had used illegal tactics to defeat Sun's Java programming system. Sun had instigated costly antitrust actions against Microsoft by U.S. and European government agencies.

So what were these killer apes doing onstage together, not quite embracing—in fact, still at arm's length—but with right hands joined, and left hands on forearm and shoulder? Why were they slapping backs and laughing in a joint display of nervous nonaggression? Why were they bantering about their common roots as Michigan-raised, Harvard-educated sons of auto industry execs? Why were they swapping Detroit Red Wings jerseys? ("Friendship and partnership starts on the ice, I guess," said Ballmer, holding up his jersey, apparently unaware of professional hockey's doubtful value as a model for peaceable behavior.)

The two companies had reached a settlement, with Microsoft forking over almost $2 billion in compensation and both companies agreeing to terms for future patent sharing. Neither side actually said, "I'm sorry," though McNealy promised to "try to be good" about not taking further jabs at Microsoft.

The motive for this reconciliation was twofold. Both companies faced common enemies in the Linux operating system and IBM. They were also under pressure from customers, shareholders, and stock analysts, who felt that their bitter rivalry was a loud and possibly dangerous distraction.

This is *chimpanzee reconciliation lesson number three:* Unreconciled conflicts take a toll on everyone. When high-ranking chimps fight, even bystanders suffer palpable tension. They watch anxiously, like small children listening for the nuances when Mommy and Daddy fight. They buzz among themselves about the latest round in the conflict, throwing wary glances at the combatants. These are dangerous moments. Anything could happen: a shift in the power structure, divorce, a hint that somebody is about to go postal.

The emotional contagion takes physiological form even in noncombatants, as cortisol level, heart rate, and blood pressure shoot up and stay there. The stress makes the chimps fretful; they scratch and fidget and pluck at themselves. (Humans do the same thing when a fight breaks out at a meeting. The bystanders shift in their seats, straighten their hair, pat down their clothes.) The entire group waits anxiously for the moment of reconciliation and dreads the possibility of renewed conflict. A public show of making peace actually relieves the physiological effects of conflict. The entire group exhales.

In fact, chimpanzees are so disturbed by unreconciled conflicts that individuals sometimes take the considerable risk of stepping between warring parties to broker the peace. The first time de Waal witnessed such a Jimmy Carter moment, the hero was a lumbering old female named Mama with "an enquiring and all-comprehending" gaze. She broke up a fight between two warring males by embracing them, one in each arm. Another time she went up to a screaming male and put her finger in his mouth, a gesture of reassurance among chimpanzees. Then she turned to his rival and called him over for a kiss, after which the two combatants embraced.

Corporate managers, who spend an estimated 42 percent of their time mediating workplace disputes, will understand the delicacy of the moment. "These males are very tense and they're dominant and strong and aggressive," said de Waal. "So to step in and bring them together is a risky business. To me, it means that she cares about relationships in her community. Chimpanzees have something like 'community concern.' They live in a group and they have to get along, and their life is going to be better if their community is better. That's the selfish motive. But this is also the basis for our moral systems: Our life will be better if our community functions better."

When CEO Andy Grove went ballistic at Intel, the senior executives who stopped the fight were taking the same risks as Mama, for the same purpose. Likewise, when the bitter rivalry between Sun and Microsoft had dragged on too long, it was McNealy's wife who started the reconciliation process, by inviting Ballmer's wife to her house for a visit. (During an interview on CNET, the two prideful male adversaries floundered around the question of how they'd actually gotten together.

Ballmer: "We have some mutual friends." McNealy: "We had a joint board member." Ballmer: "It was actually Scott's wife." McNealy: "Okay, we'll blame her.")

FORGIVENESS IN SPITE OF OURSELVES

No one has studied why a good reconciliation has such a potent, visceral effect on all parties. What's going on in the brain? What's going on in the gut? Delivered properly, the words "I'm sorry" often seem to do much more than merely relieve the stress about further conflict. They produce an almost miraculous emotional transformation.

For example, after the accident in which the nuclear submarine USS *Greenville* sank a Japanese fishing vessel, Scott Waddle, the submarine commander, apologized in person to the families of nine lost Japanese crew members. One witness reported later that Waddle bowed to the families, "and they saw his tears hit the floor. It had a profound effect. Kazuo Nakata . . . whose son is missing at sea, did not understand much English, but he said . . . he understood two of Waddle's words very clearly: . . . 'very sorry.' . . . At that moment his anger suddenly dissipated."

Vanderbilt University law professor Erin Ann O'Hara suggests that this transformation takes place almost "outside the will of the victim." With some consternation, she describes an afternoon spent brooding in anger over a transgression by her husband. At home that evening, as she was just beginning the outraged speech she had been rehearsing, her husband abjectly apologized. Her anger vanished so quickly O'Hara found herself protesting that she "wasn't finished being angry" yet. But her body told her otherwise.

A sincere apology often involves a touch or an embrace, and the renewal of damaged or broken social bonds. So, does it trigger the release of oxytocin, which lowers heart rate, respiration, and blood pressure, and also induces feelings of warmth and affection? Is the effect of an apology more profound because it represents such a dramatic swing from fight-or-flight readiness to tend-and-befriend relaxation? Unfortunately, these are blank areas in scientific knowledge, partly because it is difficult to re-create the experience of injury and reconciliation in a laboratory context—and partly because, until chimpanzees clued us in, no one really thought reconciliation mattered.

Chimpanzee reconciliation lesson number four holds that the stronger the bond between the contending parties, the deeper the yearning for reconciliation. Chimps and some other primate species are far more likely to reconcile with their social allies and kin than with mere acquaintances. Surprisingly, male chimps reconcile more often than females. This may be because they give one another more cause for apology. It may also reflect the extreme importance of social alliances in the endless dominance struggles that characterize male life. They apologize because they need each other's help in the conflicts ahead. Indeed, they sometimes make preemptive conciliatory gestures to reduce tensions in advance of a potential conflict.

At Yerkes National Primate Research Center, staffers often serve a meal in two or three bundles, to study how friendship and rank affect the tricky business of sharing food among the twenty or so chimpanzees in a compound. Feeding time is thus fraught with anxiety. Much as two humans declare their good intentions with a handshake—a literal clasping of weapon hands—and an interlude of small talk, the chimps also make obvious gestures of nonaggression. Bjorn, the alpha, allows Socko, his chief rival, to gnaw on the back of his wrist, and Socko lets Bjorn mock-bite his shoulder. They slap each other on the back affectionately and dance up and down. The three top males join together in a huddle. They are so intent on defusing the tension of the moment that they sometimes miss the first serving of food and do not seem to care.

It's like Wall Street brokerage associates getting drunk together and pledging mutual nonaggression during the tense days just before annual bonus time.

HUMAN REFINEMENTS

To be sure, our reconciliation behavior also differs from that of other animals. We follow rules that seem to be unique to our species. We're much fussier than chimpanzees, for instance, about what constitutes a sincere apology, properly delivered. So while it is important to pay attention to the biological roots of our reconciliation behavior, it is also crucial to observe a few human niceties:

- *You need to say the words.* And with certain exceptions, it's better if the words include a direct admission of personal responsibility for the transgression. It isn't enough to say, "I'm sorry it happened," as Pete Rose discovered, or "Mistakes were made." You need to say, "I'm sorry I did it." In fact, Rose actually got the wording right on ABC's *Primetime Thursday* news show: "I am terribly sorry for my actions and for my bad judgment in ever wagering on baseball, and I deeply regret waiting so many years to come clean." But the nonapology in the book, and his air of truculence, disheartened even those who yearned to welcome Rose back into the baseball fold.

People may hesitate to say the words because they fear that an apology is legally tantamount to a guilty plea. This is particularly true in the United States, where an apology to an injured party may later turn up in court as evidence of guilt. Hence even Japanese corporations steeped in a national tradition of apology urge employees going off to work in the United States to avoid apologizing, for instance, after a car accident.

Under U.S. law, an apology is protected from use as evidence in only one circumstance—when it is directly linked to a monetary offer to settle a claim. Daniel W. Shuman, a professor at Southern Methodist University School of Law, complains that this rule encourages only those apologies that are "least therapeutic or sincere." The structure of compensation for lawyers, who typically take a percentage of any legal settlement, also discourages apologies. One-third of "I'm sorry" will not buy the lawyer a tuna sandwich for lunch.

On the other hand, many states have recently passed laws making an apology or a statement of sympathy or benevolence inadmissible as evidence against the defendant, in at least some circumstances. And despite the legal risk, some scholars and insurance companies now argue that the redeeming character of a good apology may justify it on purely practical grounds. Even if it helps establish a defendant's guilt, legal scholars point out that many civil and criminal disputes "are not as concerned with liability as with how much is due." And on that question, victims and juries often accept an apology as grounds for reducing the size of the damage award.

■ *You need to say the words at the right time.* Fourteen years is too long a wait. But when an apology comes too soon, it can seem reflexive and insincere. Nicholas Tavuchis, author of *Mea Culpa: A Sociology of Apology and Reconciliation,* writes that there is "a tender moment following an offense which, if hastily foreshortened or heedlessly prolonged, is likely to harden hearts rather than allow for a salutary stirring of sorrow and forgiveness."

The timing of preemptive conciliatory acts is also critical. You can't wait until trouble is brewing before you start to act nice. People need to feel that you are being conciliatory because you value the relationship over the long term and not just so that you can use them for a momentary cause. Though it goes largely unspoken, people are acutely sensitive to this rule of the decent interval.

■ *The apology can't be too obviously self-serving.* Brothers Bob and Harvey Weinstein built their careers as movie producers at Miramax on a pattern of wild, unpredictable abuse. Harvey "tore phones out of walls and hurled them. He slammed doors and overturned tables," according to Peter Biskind, author of *Down and Dirty Pictures.* "Almost anything within reach could become a weapon—ashtrays, books, and tapes, the framed family photographs . . ."

One ex-employee recalled: "Their M.O. from the time I've known them was to stomp on you, then help you up, brush you off, and apologize." Biskind adds: "You could have produced a small film on the money Miramax must have spent on Harvey-didn't-mean-it flowers." But the manipulative character of both the stomping and the flower giving eventually became evident to everyone. It's hard to accept an apology as the mark of a genuine change in attitude when the culprit goes out and does the same thing even more viciously the next day.

■ *It has to be voluntary.* In a study of preschoolers, opponents renewed their relationship only 8 percent of the time after a teacher told them to make peace, versus 35 percent of the time when they did it on their own initiative. It doesn't get any better as we grow older. To the chagrin of managers trying to broker a cease-fire, the warring parties don't generally perform a forced apology, or accept one, with grace. But animal behavior suggests at least two hopeful possibilities:

In a study of wild baboons, researchers recorded the vocal signals of individual animals who were familiar to one another. Then, after a fight between Freddy and Sam, say, the researcher would use a hidden speaker to play back Sam's grunt, the baboon equivalent of an apology. Freddy would almost always accept the recorded grunt as genuine, and the two adversaries would soon move closer together and begin to reconcile.

Humans are no less susceptible to suggestion. Romances spring to life when friends merely suggest that one party has a crush on the other. Likewise, reconciliations can begin when a boss or a colleague tells one party to a dispute that his rival "feels terrible about what happened; he just can't figure out how to say it."

Conciliatory behavior also seems to be contagious. Rhesus macaque monkeys aren't very good at reconciliation. But when they are reared together with stumptail macaques, a species with a strong tendency to reconcile, the rhesus macaques learn to become skillful peacemakers, too. Likewise, when the boss makes a practice of apologizing for transgressions, subordinates are also more likely to seek the conciliatory path.

At one institution, a subordinate disobeyed a direct order because of what she claimed was a misunderstanding. Then she missed the next big staff meeting to avoid a confrontation.

Her manager opted "to be big, by making myself small." At the meeting, she noted that the subordinate wasn't present to defend herself, and in any case, she didn't want to rehash their conflicting accounts. "Instead of pursuing the aggressive path, I said, 'Okay, I could be the one who's wrong, let's move forward. I'm trying to initiate reconciliation.' I actually said that. This is somebody I need to work with in the future."

It would be gratifying to say that the subordinate was moved to apologize. This is, however, the real world: She merely deigned to accept her boss's peace offering. But her boss was satisfied: "I think there's a long-term payoff. I showed them I'm willing to cooperate. I can be modest. I don't have to be right. I said it in public, at the meeting, so there would be no secret. I thought this would be a good way to strengthen my position. Afterward, a friend of mine said, 'I have to hand it to you, you really put yourself out there.' "

NEVER APOLOGIZE, NEVER EXPLAIN?

There are, to be sure, circumstances where the miraculous words "I'm sorry" can make us weaker, not stronger. Subordinates may take a boss's apology as a signal that they can flout the rules. An employee who apologizes to outsiders may find himself in trouble for betraying the company culture, or for exposing management to criticism and blame. A gang member who apologizes would surely lose credibility in the macho culture of the street. An Osama bin Laden or a Timothy McVeigh would gain nothing by apologizing to the families of his victims, because some crimes are too heinous to be redeemed by an apology.

Otherwise, how many people can you name who have lost standing because of a full apology? Starting with the ridiculous, did the actor Hugh Grant suffer when he apologized on national television to his actress girlfriend Liz Hurley after having been arrested in flagrante delicto with a Los Angeles prostitute? On the contrary, he stammered and got forgiven, at least by the general public, with the result that his career flourished.

Or to take a far more profound example: Did former U.S. national security adviser Richard Clarke lose credibility when he turned to the families of the victims of the September 11 terrorist attacks and apologized to them for his own errors and those of his government in failing to prevent the tragedy?

It's easy, on the other hand, to think of people who have been permanently tainted by their unwillingness to make a forthright statement of apology and regret. The short list includes Richard Nixon, Bill Clinton, George W. Bush (who refused even to acknowledge mistakes during a press conference shortly after the Clarke apology), Pete Rose, and Rush Limbaugh. An apology acknowledges that we have violated the standards of acceptable social behavior and regret the harm we have caused. It makes us stronger because it gains us readmission to the human community.

The slogan "Never apologize, never explain" aims, by contrast, to put the individual above the community and beyond ordinary laws or social norms. It's an attempt to "banish sorrow and forgiveness from human affairs," according to sociologist Nicholas Tavuchis, and to transform the guilty parties "into either gods or robots." But gods and robots evoke resentment, especially when they do harm to flesh-and-blood humans.

The failure to apologize thus often leads to what biologists call "moralistic aggression," including costly and acrimonious lawsuits, workplace sabotage, and other retaliatory behaviors. Douglas Yarn, a law professor at Georgia State University, helps mediate disputes in the statewide university system. When a conflict does not reach a satisfactory resolution, he says, it's common to see an aggrieved party respond with rumor-mongering, badmouthing the leadership, faction forming, and general noncooperation—in other words, all the normal behaviors of the average dysfunctional university department. An apology often goes a long way toward putting these things right.

The ineluctable appeal of an apology can, on the other hand, also make it dangerous. In a paper about the role of apology in modern legal disputes, Yarn and coauthor Erin Ann O'Hara suggest that our evolved propensity to reconcile can make us vulnerable to people who use an apology opportunistically.

It goes back to O'Hara's observation that the collapse of anger produced by the words "I'm sorry" seems to happen outside the will of the aggrieved party. "An organization—whether government, corporation, or other association—can take advantage of victims' predisposition to forgive in order to minimize its liabilities. . . . An institution that wishes to exploit victims' cognitive and emotional structures will send its most empathic employee or member to apologize to the victim."

Yarn and O'Hara thus argue for certain limits on the use of apology. For instance, a company should not be free to use an apology for the purpose of reducing damages unless that apology is also available to the plaintiff for the purpose of establishing guilt. Likewise, a liability release should be enforceable only if the people signing it have had an attorney to help them deal objectively with the natural emotional response elicited by a sincere apology.

Yarn and O'Hara seem to put their trust mainly in another evolved predisposition: Humans are highly skilled at detecting cheaters. It is an essential inheritance from our tribal past. We use words, facial expressions, body language, and long-term engagement as indicators of sincerity. The formal process of mediation provides a useful forum for cheater detection, because both sides are free to say what they really think. Even if the mediation fails and they end up in court, the rules generally prevent participants from using each other's words in evidence. So the defendant has a safe haven for delivering an apology, and the plaintiff has the opportunity to look closely at the defendant and decide if the apology is genuine.

The companies that use this process most effectively are naturally the ones that seem to mean what they say. For instance, the Toro Company, a Minnesota manufacturer, faces about 125 personal injury claims a year, typically from people who have put a hand down a snow blower chute or gotten injured in a lawn mower accident. Toro used to follow standard "deny and defend" corporate practice against all claims, and its product liability costs soared. But in 1991, it switched to a more conciliatory approach.

Now, when the company first hears about an accident, it immediately sends one of its two female paralegals to visit the family, along with a company engineer to inspect the machine and the scene of the accident. They always begin by expressing regret, roughly on these lines: "Setting aside the question of who's at fault, we want you to know that we feel terrible that this happened. We're extremely unhappy when people have accidents using our products. We're going to do our best to resolve this thing and make sure it doesn't happen again."

In one case, Toro employee Drew Byers found the victim of a lawn mower accident cooped up in a miserable apartment, in a full body cast, in midsummer. "He was going to be lying around for a couple of months, and not looking very comfortable," says Byers, who is the company's manager of "product integrity" (a notable variation on "product liability"). Byers bought an air conditioner and had it installed in the apartment next day.

Was this gesture manipulative? Or was it a sincere expression of the company's sympathy? The plaintiff believed it was sincere. During the mediation process, his lawyer warned him not to take a settlement offer because "this is just a big fat corporation out to screw you."

The plaintiff then took Byers aside in private and asked if the proposed settlement, for around $100,000, was really the best the company could offer. Then he went back to his lawyer and said, "Don't you tell me this guy is trying to take advantage of me, because when I was lying in a full body cast in ninety-eight-degree heat, he came out and took care of me."

In another case, Corey Soles, a seventeen-year-old, came too close to an embankment while operating one of Toro's zero-radius-turn ride-on lawn mowers in northern Florida. The machine rolled over and landed on him, breaking his neck. His parents blamed Toro for their son's death, because the lawn mower had not been equipped with a roll bar. But after Toro deployed its policy of regret, condolences, reconciliation, and face-to-face engagement, even Corey's mother, Debbie Soles-Smith, described the company as "awesome," "very sincere," and "really supportive."

Listening to the lawyers in the case now, it's hard to tell which side they represent. Mike Olivella describes Corey Soles as "a model student, straight A's, any parent would have been proud to have him for a child. He was such a popular kid that two thousand people showed up for his funeral." Olivella represents Toro.

Norwood Wilner, the attorney for the parents, describes his adversaries in turn as "such gentlemen, so pleasant," and adds that both he and the boy's parents were impressed with "how much they cared about Corey Soles and how serious they were about making sure this didn't happen again."

Wilner is an affable, aggressive sort with a long background in product liability suits against companies with a policy of denial and harassment. Such companies typically try to wear down the plaintiff with endless motions and painful depositions. Wilner's job as a plaintiff's attorney is to hand back a heavy dose of moralistic aggression.

But what Toro communicated, he says, was a willingness "to face up with it and deal with it honestly and pay the claims that are legitimate." During mediation, the company sent down an engineer. "This guy Bud was as good a guy as you can imagine, a big old guy from Minnesota, not just a figurehead, a guy who was involved in the design of these things. It wasn't just like they sent a pretty face down to paste things over. They allowed me to question him about why the machine didn't have a roll bar, and his honesty just floored me." Toro's zero-turn-radius lawn mower already cost almost $6,000, a critical price point in a competitive market. So Toro offered a roll bar, but only as a $700 option.

As part of the settlement, Corey Soles' parents were seeking an agreement to install the roll bars on all such lawn mowers. Toro agreed only to consider the roll bar question, and also to pay the family $500,000 in compensation. Wilner says his clients were satisfied. "We ended up just where we were going to end up, anyway, and they felt good about it. Justice was done."

A few months later, Toro announced that it would install roll bars on all such mowers in the future. It also made a retrofit roll bar available to past customers at cost.

For Toro, one benefit of its conciliatory approach is that people feel good about working there. Customers are also reassured by the company's willingness to stand behind its products. Mike Olivella says his firm now represents a dozen or so other major corporations following the Toro example. But all of them refuse to be named out of fear that going public might cause an increase in liability claims.

According to Toro's Drew Byers, the conciliatory approach has had little effect on the number of claims. It has, however, cut the time to reach a settlement from two years to nine months. The average cost of a claim, including lawyers' fees, has dropped from $115,620 in 1991 to $35,000 in 2004. In addition, the annual cost of the company's liability insurance dropped by $1.8 million, before Toro opted to self-insure. By 1999, an intuitive understanding of the natural history of "I'm sorry" had thus saved the company an estimated $75 million.

Informed of these numbers, Wilner, the product liability lawyer, expresses delight: "The cry of corporate America today is that they're getting killed in litigation. But what keeps a lot of the defense firms in business is this policy of driving the cost of litigation up. The more paper you file, the longer it takes, the harder it is, the more the plaintiff has to suffer, that's your job. And then the companies complain, 'Why am I paying all these bills?' Well, the people who work for you are running up the cost. And here's this company that does it a little differently and they save $75 million. I *love* it."

Cleanerfish are the beauticians and personal dressers of the marine world, and also among its master apologists. These little fish, typically with a black stripe down their flanks as a kind of livery of their profession, inhabit small territories, or cleaning stations, on coral reefs. Other fish drop in, sometimes as often as thirty times a day, for bouts of personal grooming. The cleanerfish inspects the client fish and picks off copepods, isopods, flatworms, and other parasites.

Sometimes the groomer gets a little overzealous, particularly if a client doesn't have enough parasites to keep it decently fed, and it nips a morsel of the client's own flesh. The client reacts angrily. So cleanerfish have evolved a gesture of reconciliation. They hover just over the client's back (a safe distance from the teeth) and flutter pectoral and pelvic fins to deliver a soothing massage.

This small gesture works so well to soothe ruffled scales that when business is slow, cleanerfish hang out in front of the shop and give little come-on massages to get potential clients to take a break from the daily hustle.

10

MAKING FACES
A Field Guide to Facial Expression

He who will . . . watch the countenance of a monkey when insulted, and when fondled by his keeper, will be forced to admit that the movements of their features and their gestures are almost as expressive as those of man.

—CHARLES DARWIN

Look at your face in the mirror. Smile. Grimace. Be frightened.

You have spent your life with this specter, this enchanting, distressing window onto your soul, and you probably operate under the delusion that its familiar expressive quirks, its fetching dimples, its wry upturning at one corner of the mouth are—*ah, love!*—entirely individual and also largely under your control.

In fact, your facial expressions are animal behaviors.

The human face uses forty-three muscles to create ten thousand visible facial configurations, of which perhaps three thousand are meaningful. And they mean much the same things in any company, any culture, any country in the world. Whether you are on the factory floor of a BMW assembly plant or among the most remote mountain tribes of Irian Jaya, the same shifting muscle movements reveal what you are feeling. They reveal your feelings even when you least want them to. If you pay attention, some researchers say, they will reveal the true feelings of everyone around you, too.

DARWIN'S DOG

The biological roots of our facial expressions thread back across millions of years and through multiple species, and anyone who watches other mammals, birds, or even fish soon comes to appreciate it. No other animal behavior is so richly expressive as the human face, but animal facial expressions and body language still unmistakably communicate emotions, from wrath to affection.

The dim foreshadowing of our own emotional life is one of the things that most delight people at zoos, or at home with their pets. "I formerly possessed a large dog, who, like every other dog, was much pleased to go out walking," Charles Darwin wrote in his pioneering book *The Expression of the Emotions in Man and Animals.* "He showed his pleasure by trotting gravely before me with high steps, head much raised, moderately erected ears, and tail carried aloft but not stiffly. Not far from my house a path branches off to the right, leading to the hot-house, which I used often to visit for a few moments. . . . This was always a great disappointment to the dog . . . and the instantaneous and complete change of expression which came over him as soon as my body swerved in the least towards the path (and I sometimes tried this as an experiment) was laughable. His look of dejection was known to every member of the family and was called his 'hot-house face.' This consisted in the head drooping much, the whole body sinking a little and remaining motionless; the ears and tail falling suddenly down . . . the eyes became much changed in appearance, and I fancied that they looked less bright. His aspect was that of piteous, hopeless dejection; and it was, as I have said, laughable, as the cause was so slight."

Darwin argued that the resemblance to human expression was no accident, but the result of our having evolved together through eons of natural selection. We now have ample evidence that he was right. For instance, we know that primates and rodents evolved from the same ancestors before they began to go their separate ways roughly sixty-five million years ago. And when the person in the next cubicle pops up like a prairie dog and twists her mouth sideways in consternation at something you just said, the facial gesture connects her directly back to the biological heritage we share with real rodents. You may thus, in all scientific accuracy, call her a "rat-faced little busybody," though only if you are also willing to apply the same description to yourself. University of Utah scientists recently demonstrated that this human facial movement, among others, is under the control of the same three or four genes that enable a house mouse to roll its eyeballs sideways, wiggle its whiskers, pull back its ears, or blink its eyelids. The so-called Hox genes guide embryonic development of the nerves in the hindbrain that direct these muscle movements in all mammals. Thus Disney employees give good Mickey in more literal ways than they imagine.

Primatologists naturally see a much closer connection to human facial expressions. Apes and monkeys do not shed tears in sadness or express disgust as we do. But our smile comes from their fear grin, the silent teeth-baring display, which they use to express submission and appeasement. Our laughter comes from their "play face," the same relaxed, openmouthed expression dogs also use when they want a friendly tussle.

DON'T BLINK

The tendency is to think that reading faces is easy, as natural as figuring out what your dog wants. Salespeople in particular often fall for popular self-improvement folklore about the face: "If he looks up and to the left, he must be lying." The puzzling truth is that most of us are dismal at reading facial expressions, particularly the facial expressions of strangers. Often we don't even bother to look. And when we do look, we are frequently clueless about what facial expressions mean. In research studies, police officers and even CIA polygraphists typically do no better than guesswork at separating straight talk from lies. Medical professionals do worse than some amateurs at recognizing pain on a patient's face. And corporate managers often seem to be oblivious to all facial expressions. This is like working as a zookeeper and not understanding that when the leopard puts his ears back he is thinking about inviting you for dinner. A refresher course in reading the human face ought to be mandatory before any manager ever steps into the cage.

In a classroom in San Francisco one morning, a mild, slightly stooped psychology professor named Paul Ekman slipped a CD into a computer. His students watched as a series of ordinary faces passed across the screen. Each face displayed an emotion for one second, and the task was to identify it as anger, disgust, fear, surprise, sadness, happiness, or contempt. The brevity of the display was true to life. Most full facial expressions last just 0.5 to 2.5 seconds.

In any case, it should have been easy. Ekman's students were themselves teachers, with years of experience peering into the faces of their own students for hints of interest, understanding, or (lately) homicidal rage. But like most groups, this one got half the fourteen test samples wrong, and mostly for predictable reasons.

We often mistake anger for disgust, Ekman told them afterward, because both involve lowering the brow. He showed a photo of a face in anger and pointed out a secondary distinguishing characteristic, the way the lips press together to the point of going white. Then he showed another face, also with lowered brow, but with the telltale sign of disgust, a scrunched-up nose. Once pointed out, the difference was hard to miss.

"Why is that surprise and not fear?" he asked of the next photo. In both expressions the eyebrows shoot up and the mouth opens. "Because the lips are relaxed. My mother used to call this 'catching flies.' " In fear, the mouth would tighten and pull back toward the ears.

As his students became more adept, Ekman shortened the exposure time. "Anger," someone called out.

"I didn't see anything," a voice protested.

"Don't blink," Ekman advised.

It was in some ways a tantalizing exercise. Ekman's students typically learn to read facial expressions far more accurately with as little as an hour of training. In one experiment, Ekman ran his students through a twenty-three-minute CD-based training program. Afterward, he showed them news footage of public figures, and the trained students accurately read the emotions being expressed 50 percent of the time. The spy Kim Philby showed distress in his last interview before fleeing Britain. The O. J. Simpson career houseguest Brian "Kato" Kaelin concealed anger at prosecutor Marcia Clark.

Being right half the time is still pretty dismal, and not nearly good enough to make you dead certain, say, that your sales rep is about to blow the Hyundai contract. On the other hand, untrained subjects got the emotion right just 10 percent of the time.

So what's going on here? Why should we need training anyway? Facial expressions are part of our biological heritage. Shouldn't a knack for reading them be second nature, too?

BUZZ LIGHTYEAR'S SNEER

Until Paul Ekman came along, the emotions passing across our faces were about as difficult to measure or analyze as gusts on the surface of the sea. But in the 1960s, Ekman and Wallace Friesen, both psychologists at the University of California at San Francisco, developed a scientific way to recognize and interpret every possible human facial expression.

Their Facial Action Coding System, or FACS, originally a five hundred-page manual, now a CD-ROM, breaks each facial expression down into its component muscle movements, or "action units," and it has become the essential tool in the science of faces. The system has also given Ekman, whose books include *Unmasking the Face* (with Friesen, 1975), *Telling Lies* (2001), and *Emotions Revealed* (2003), the basis for exploring how the face affects every aspect of our lives, from the bonding between mother and child to the expression on a suicide bomber about to flip the switch.

Some critics have argued that Ekman overstates the link between facial expressions and underlying emotions. Zoologist and *Naked Ape* author Desmond Morris, on the other hand, credits him with bringing "scientific rigor to a topic all too often treated as an anecdotal game."

Ekman's research has earned him a weirdly diverse following. The Dalai Lama is providing seed money for him to develop a prototype course on "cultivating emotional balance." At the same time, counterintelligence agencies routinely hire him to teach the nuances of facial expression for use in interrogating suspected al-Qaeda terrorists. (They can get awfully spooky about a subject that is literally in front of our faces. When I was researching Ekman's work, the U.S. Office of the National Counterintelligence Executive agreed to let me attend one of his training sessions only if they could review the finished work for "security editing." I declined.) As if this dichotomy were not enough, Ekman's work has also enabled Buzz Lightyear in *Toy Story,* and a generation of cartoon characters since then, to arch their eyebrows and otherwise make faces like real people. Pete Docter, who directed *Monsters, Inc.* for Pixar Animation Studios, says the FACS atlas "made us aware of things" no one had noticed before. It helped "pinpoint the little things that might turn out to be the essence" of, say, a sneer.

SOPHISTICATED EXPRESSIONS

At his home in the Oakland hills one afternoon, Ekman, in slacks and a rumpled, colorful sweater, was professorial and self-deprecating. He's in his seventies, with thin gray hair, white sideburns, and rimless eyeglasses. His characteristic facial expression is contentment, his eyes bright, the outer corners of his lips turned slightly upward, though with hints of peevishness. A few minutes earlier, when he'd arrived home to find me already waiting in the driveway, something in the way he glanced out the window of his car told me that I'd taken his parking space, and I promptly moved my car. We went down into the living room, where a glass wall looks out across San Francisco Bay to the Golden Gate Bridge. At that moment everything was shrouded in fog.

Ekman soon warmed to his topic. He came to his work, he said, with a natural advantage. "My mother always used to say, 'Stop making those faces, you'll freeze your face.' Less than 2 percent of the population can do this," he said, and proceeded to make left and right eyebrows pop up and down in alternation. Having a highly mobile face (he can also wiggle his ears independently) is the product of genetic predisposition.

"When I was a kid in Newark, New Jersey," he said, "the ultimate sophisticated expression was like this." He tilted his head to one side and cocked the outer side of his left eyebrow up. (In FACS, this is a U2L, for unilateral outer left frontalis muscle.) Then the inner corner of his right eyebrow came in and slightly down (a U4R, for inner right corrugator muscle). The result was a picture of smirking disbelief. "It's a mechanical system. Everything you see is muscles attached to skin, and they're pulling the skin around. That's what produces expressions."

Ekman got his start in facial expressions as a young psychologist when the Department of Defense gave him a grant to figure out which aspects of nonverbal communication are cultural and which are universal. He soon found himself deep in Papua New Guinea with an isolated "Stone Age" tribe. His test subjects had never been outside their own culture, nor ever seen a movie or photograph. But they raised their eyebrows in surprise, bunched up the muscles around their eyes in delight, and glared in anger much as we do. In a series of experiments over two trips, Ekman concluded that facial expressions are universal and biologically determined.

At the time, most students of human behavior still believed in the primacy of culture over biology, and the anthropologist Margaret Mead, among others, scorned Ekman. The Department of Defense got taken to task, he recalls, for "wasting the taxpayers' money studying the facial expressions of savages."

Undeterred, Ekman joined up with Wallace Friesen, and the two of them spent seven years, in essence, making faces at each other. Anatomists had long since peeled back the skin and recorded the musculature of the face. But no one had charted what those muscles do. Ekman and Friesen catalogued every possible movement by learning to voluntarily move each muscle in their own faces.

HONEST SIGNALING

Ekman came to believe that our facial expressions are not merely biological but also hot-wired to our emotions, revealing anger or fear even when we least want them to. It takes just two hundred milliseconds (less than a quarter second) for a stimulus to produce an expression on the human face. The conscious mind needs more than twice as long to recognize that the emotion is even being felt, much less displayed.

Ekman argues that this hot-wiring evolved because facial expressions originally served for "honest signaling." Our tribal ancestors lived in small groups of friends and relatives, where putting on a false face to lie or cheat carried the risk of being cast out into a hostile world. When he lived among tribal people in Papua New Guinea, Ekman said, "everything was based on cooperation. If they were ostracized, they were basically dead."

No form of honest signaling matters more than smiling. It's so important that humans have evolved anatomically distinct varieties to communicate everything from cooperativeness or accommodation to polite tolerance. For instance, former President Bill Clinton has a characteristic "grin-and-bear-it smile," according to Ekman, with lips pressed together, mouth corners up, and chin bunched, "a 12-17-24, the expression you give the dentist when he says you have to have a root canal." The flirtatious smile, on the other hand, involves looking away, lifting the cheek muscles in a "felt smile," then glancing back and away again, just long enough to be noticed. A felt smile involves crinkling around the eyes, which most people cannot fake. Humans have evolved to spot a smile more readily than other expressions, even at the far end of a football field, apparently because it's a way to judge if people coming toward us bring good news or woe.

Misunderstanding these smiles can mean calamity. For instance, a wife's merely polite smile, with all the movement in the lips and none around the eyes, may foretell hurled table lamps. And yet even within the family, facial expressions often elude us. His own children, said Ekman, never figured out that when their mother asked, "Where were you last night?" with her eyebrows down, this was a sure sign she didn't know the answer. Bluffing was still an option. But if she asked with her eyebrows up, all hope was lost.

We are even worse at reading the facial expressions of coworkers, though the calamitous possibilities can at times be tragic. After a workplace shooting or a suicide, or when a valued employee quits and goes elsewhere, colleagues frequently report that they saw no warning signals. But why not? If our faces have evolved to send such nuanced messages, why aren't we better at reading them?

THE LANGUAGE BARRIER

It may be that we are victims of our own recent evolution. Some geneticists date the origin of language back as little as fifty thousand years, and the richness of words actually seems to distract us from the older medium of faces. In one study, stroke victims and others whose brain damage made them less attentive to speech were far better at focusing on facial expression. They picked out the liars 73 percent of the time, an accuracy level rivaled, in Ekman's testing experience, only by the professionally risk-sensitive (and word-averse) agents of the U.S. Secret Service.

By coincidence, Helen Starkweather, the researcher who helped me when I was checking facts for this chapter, was born with profound hearing impairment and has learned to compensate by paying careful attention to facial expressions and body language. "Often, I will accidentally think that someone has spoken," she told me, "when in fact, their face spoke so clearly, albeit silently, that I mistook their expression for a sentence. Which can be embarrassing when people want to hide what they're thinking. You have to dance around that and sort of 'not know.' "

People may well not want to know the things that facial expressions make nakedly apparent in the absence of language. A friend of Starkweather's, also hard of hearing, was working on a new interior for a client, a busy saloon owned by a husband and wife: "One day I noticed an employee look at the husband with hope and longing on her face. He saw her, smiled, and drew in a deep breath. The wife arrived later and glared at her husband, I mean hate was in those eyes. Later, the wife left, and I had to go, too. Unfortunately, I forgot to have my work order signed. So I went back and walked in on a very amorous situation. On a pool table. I beat a hasty retreat and nearly knocked the wife flat as she was heading back in. I finished the work for the new owner soon after the divorce and haven't forgotten to have my work order signed ever since."

Discretion, and the clamor of modern life, encourage most people to avoid paying close attention to faces. We may "learn to be bad" at reading faces, according to Ekman, because we inhabit a drastically different world from the one in which our facial expressions evolved. Instead of spending a lifetime among the familiar faces of a small tribe, we now see hundreds of new faces daily, and we have learned to cope with overcrowding by not looking into the faces of strangers or otherwise intruding on their privacy.

Modern life also encourages people to avoid too much "honest signaling" with their own facial expressions. Putting on the poker face is a way to avoid contact, preserve privacy and anonymity, and get on with their lives. "When I go into my office and say, 'How are you today?' to my secretary, I don't want to know if she had a big fight with her spouse last night," Ekman admitted. "I want to know if she can do her job. And if I say, 'Gee, Wanda, you seem upset today,' and it spills out, then I have to deal with it. And that's not why I'm there. I'm there to do my job."

TURN OFF THE SOUND

We all learned in grade school that it can be hazardous to stop listening, because that's when the teacher always calls on you. But take a calculated risk at your next meeting and shut off the sound. Try practicing first by watching a talk show with a high level of emotion, but with the mute button on. What you lose in verbal content, you will gain, and then some, in greater understanding of nonverbal cues.

"My surroundings during meetings are people's faces," says a liquor company salesman with profound hearing loss. "To keep awake, I look around to determine whether or not I should applaud, laugh, or be genuinely amazed at how well we are—or are not—doing."

At one meeting, a past sales manager who had moved up in the company came back to visit. The current sales manager "gets up and gives us the numbers and 'how we *have* to do better, we *have* to do better than last year. We *have* to give it that extra 10 percent.' During this tirade, people's faces go from faking 'happy to be here' to sullen. Their eyes just stare and they don't blink much."

The past manager also loses his smile as he looks around the room. "He knows hate when he sees it. And he's up next. He is not happy.

"So when it's his turn, he does the best possible thing: He tells jokes. He tells us what it's like not having us with him at his new job, and how much he misses us. And he does all this sincerely. We look in his eyes and know he's speaking the truth. His walk is confident. He uses his hands and arms to convey honesty.

"His replacement, to this day, has none of these attributes. This guy's demeanor is all job title. He likes to imitate the John Wayne walk, just rolls right up, plants his feet a little wider than his shoulders, with this stern look on his face and his arms crossed. He won't meet your eye even if you walk up and shake his hand and say 'Hi!' Doesn't remember your name. Remembers the bad more often than the good. People look at him and know he's a liar, know he's not going to back them up when times are bad."

By paying closer attention to such visual cues, you may gradually discern what behaviors work when you are up against some Jabba the Hutt wannabe or some overblown John Wayne imitator—and you will have what it takes to triumph (or appease).

So how do we get past the language barrier without being deaf? How do we learn to pay attention to the facial expressions that count even when the rush of daily business distracts us? Recovering our half-lost knack for faces can be surprisingly easy. In Ekman's class on cultivating emotional balance, students quickly learned to spot the underlying emotion in facial expressions displayed on a projector screen for just one second. Then they moved on to expressions displayed for less than a fifth of a second. These "microexpressions" can be the most important back channel in a conversation, he said, because they are involuntary and reveal what isn't being said with words, and often never will be.

Ekman and Friesen first discovered microexpressions while repeatedly viewing films of a depressed woman seeking a weekend pass from a mental hospital. She appeared stable. But in slow motion, the researchers caught a flash of utter despair, the corners of her mouth yanking down, the insides of the eyebrows arching up, before the microexpression vanished again behind a smile. Some of the hospital staff apparently saw it, too, because they vetoed the weekend pass. It turned out the woman had been planning to go home to commit suicide. This sort of "leakage" of true feelings frequently occurs when people are telling lies.

Repeatedly looking at a film in slow motion or identifying microexpressions on Ekman's training CD (where a still photo of a neutral face is immediately followed by the same face displaying a microexpression) is a long way from being able to make sense of the confusing welter of expressions we meet in our everyday lives. But the students learning to pick up these microexpressions in Ekman's class seemed to feel an exhilarating sense of new possibilities. It was like rediscovering the world through that other neglected sense, the power of smell.

FACE READING FOR SEVEN BASIC EMOTIONS

Here's a quick guide to facial expressions for seven emotions commonly encountered in any workplace. Keep in mind that most unconscious expressions are brief. If someone smiles or glowers at you for longer than about 2.5 seconds, then it's a conscious effort to get you to see how he feels. That, or he's shining you on.

Also keep in mind that the photos below mostly show intense expressions. In real life, it's all about picking up the subtle, suppressed facial expressions indicating that your boss or your best account executive is struggling not to tell you that your latest big idea is just stupid. It will help to remind yourself that, in the grip of our own emotions, we tend to overlook even the most obvious evidence that other people feel differently. It may also help to brush up with Ekman's training CDs, available for $50 from www.paulekman.com.

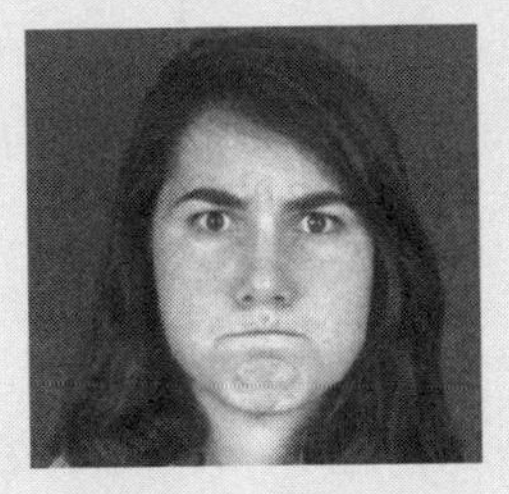

Anger: The eyebrows come down and draw together, often producing a vertical wrinkle above the nose. The lips press tight. The lower eyelids go tense, and the stare becomes hard, fixed. By themselves, these muscle movements may indicate mere concentration. But the combination of the three is unmistakably anger.

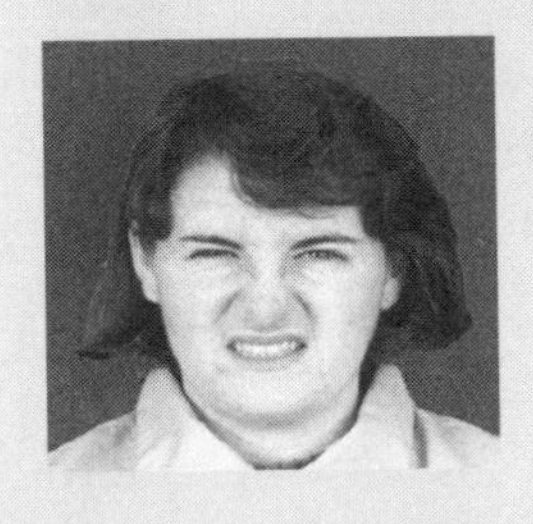

Disgust: Again, the brows come down (but not together). The nose wrinkles, the upper lip rises, the cheeks lift, and horizontal lines appear below the eyes.

Contempt: One corner of the mouth tenses and goes up. Sometimes it's just the mildest tightening of the muscles (useful for communicating, with the away side of the mouth, that the guy at the head of the table is a loser). The upper lip on one side may also rise to a full-fledged sneer or smirk.

Surprise: The eyebrows shoot up and curve, stretching the skin above the eyes. The whites of the eyes show. The mouth falls open with slackened lips.

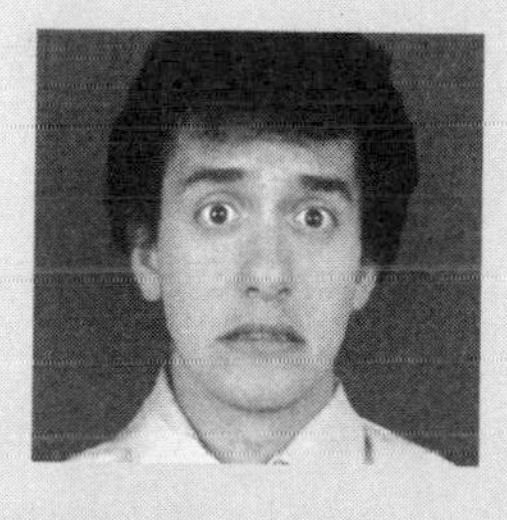

Fear: The raised eyebrows knit together. Horizontal lines appear in the center of the brow. The upper eyelids rise. The mouth opens and the lips stretch back.

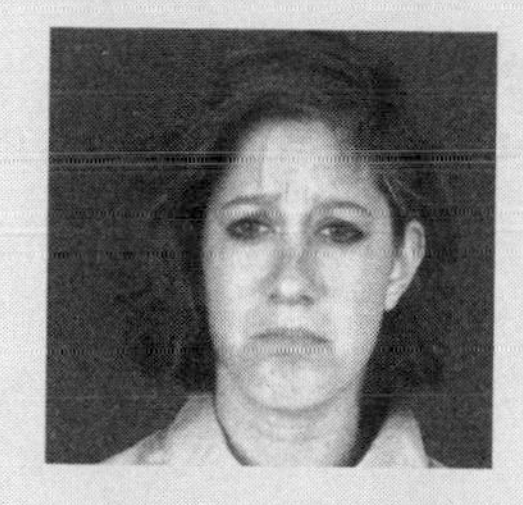

Sadness: The inner corners of the eyebrows arch up, an expression most people cannot fake. (Woody Allen and Jim Carrey are exceptions, and both use this as part of their comic personas.) The cheeks rise, and the corners of the mouth pull down, sometimes with the lower lip trembling.

Happiness, enjoyment: Not just a smile, but crow's-feet crinkling the outer corners of the eyes, with lower eyelids raised but not tense.

LEVERAGING SENSORY BANDWIDTH

Discovering the power of face reading inspires people to dream, if only about new ways to make a buck. (You thought they were going to use it to find out how Wanda really feels?) Dan Hill, founder of a Minnesota company called Sensory Logic, was working in the corporate world when he saw an article about facial action coding. At the time, he was dealing with a CEO who was "underwhelmed with focus groups," because people in these groups tend to say what they think the sponsor wants to hear. "It's almost a geographic thing," says a marketing executive for Nationwide Insurance, which became one of Sensory Logic's early clients. "People in the Midwest love anything you show them. It's hard for people being paid $60 an hour to give negative feedback."

But their faces tell the truth, according to Hill, who likes to quote Oscar Wilde: "It is only the shallow people who do not judge by appearances." Hill's approach at Sensory Logic is to get beyond traditional focus groups by combining consumer interviews with videotaping of facial expressions, plus biofeedback sensors on frown and smile muscles, and an additional sensor on a finger to measure sweat-gland activity.

In one beta test, the company asked research subjects how they felt about a dog food package. Their verbal response was about 70 percent favorable. But their unconscious facial reaction was "all negative," said Hill, and no wonder, since the dog featured on the package looked "rather rabid."

For the Federal Home Loan Mortgage Corporation, or Freddie Mac, another client, Sensory Logic tested audience reaction to a television commercial featuring an adorable little girl. The verbal response was 80 percent positive, but facial response was mostly negative. When pressed in follow-up interviews, some people admitted that having a preschooler talk about mortgages made them feel manipulated. Others just wanted her to shut up because her voice was so annoying.

Hill, who earned his Ph.D. in literature, has come to believe that words mean almost nothing. "Access to what is in your customers' heads requires an understanding of what is happening inside their bodies," he writes, and faces are better than words as a way to climb inside. He cites the familiar scientific evidence that 95 percent of all thought is unconscious. So people may not be able to tell you what they do not know they are feeling. And even if they do know, about 80 percent of communication is nonverbal, so they aren't likely to put it into words.

In particular, Hill argues, the decision to buy a product is almost entirely emotional, not rational. He quotes one estimate that 70 percent of all purchase decisions "are impulse buys that happen within the last five seconds before the decision is made." So using the faces of potential customers to understand how they react to a package or display can make or break a product. "Human bodies were designed for a lifestyle more like that of a housecat," he writes. "We respond quickly and intuitively to whatever sparks fear or happiness. We're not wired to act like a computer that checks off the variables as it considers whether an offer makes rational sense."

In Hill's view, leveraging our "sensory bandwidth," particularly by learning to pay attention to facial expressions, is the way to recognize those sparks of fear or happiness. He predicts a payoff in marketing advertising, and other areas where the goal is "to connect more fully with consumers to increase market share."

Some researchers and government agencies believe that computers using literal bandwidth may be an even better way to decipher what is going on inside people's heads, not just in focus groups but in public places. Because the Facial Action Coding System breaks facial expressions down into all their component parts, it should be a relatively straightforward business to teach computers to read faces. The thinking goes that if we could just link airport security cameras to computers programmed with FACS, we might even be able to spot would-be hijackers by a telltale glower or a microexpression of contempt.

"I'm a true believer that reading facial expressions can be done by machines, probably better than by people," says Takeo Kanade, a researcher at Carnegie Mellon University who is leading a Department of Defense–funded effort to achieve automated recognition of facial expressions. But he adds that sorting out the bugs could take twenty or thirty years.

One problem is that computer analysis goes haywire when the subject's head rotates or tilts, as the heads of people in airports routinely do. The computer also misses most microexpressions. Another problem is that people vary widely in their normal or "baseline" facial expressions.

One marker of potential trouble, for example, is the sequence of action unit 12 followed by 15—the corners of the mouth pulling down, then quickly back up—which some experts say can indicate chronic anger masked beneath a polite smile. So a computer seeking Osama bin Laden might zero in on a hard-driving businessman instead.

CULTURAL CROSS FIRE

To some critics, the whole idea that facial expressions are a direct readout of underlying emotions is flawed, entirely apart from the Orwellian prospect of having computers exploit these expressions without our consent. They argue that people may be much better at managing expressions than Ekman has suggested. "Across the world's cultures, do happy people smile? Do angry people frown?" asks Boston College psychologist James A. Russell. "Evidence on such questions is missing." Some cultures avoid *any* public display of facial expressions. In Japan, for example, anthropologist Edward Hall theorized that the traditional "emphasis on self-control, distance, and hiding inner feelings" was rooted in history: "At the time of the samurai knights and nobles, there was survival value in being able to control one's demeanor, because a samurai could legally execute anyone who displeased him or who wasn't properly respectful."

Modern-day flight attendants and airport security guards can dispatch a suspect almost as summarily, which is why seasoned travelers put on the "airport face," bland and emotionless despite endless assaults on one's patience and dignity. And if ordinary travelers can do it, why couldn't a terrorist also look serene in the sincere belief that he is about to earn his eternal reward?

For critics, the cultural factors can be at least as important as the biological ones in the business of reading faces. For instance, almost all studies show that women smile more than men, often making it their business, as one such study put it, "to neutralize tense situations and to try to rescue all parties from the talons of social awkwardness."

Some evolutionists would argue that women have adapted to the disproportionate burdens of motherhood with a biological propensity for smiling, as a way to calm the children and also hold together the fragile support network of mates, grandparents, and friends. Some feminists counter that women do the smiling simply because society sticks them with the "emotion work." (Either way, men get to act impassive, grunt, glower, and otherwise get on with their surly business of establishing social dominance over other males.)

Yale University psychologist Marianne LaFrance does not take a hard line in the evolution-versus-culture argument. She notes that gender differences in smiling vary with nationality, ethnicity, and age—and also that the differences dwindle when men and women hold equivalent rank or perform similar tasks. Women still tend to smile more in emotionally charged situations. But the dynamic of nature and nurture is, as always, complex.

"Faces are powerful instruments," says LaFrance, "but they don't stand alone. They're in a body, and the body is in a place, and the place is in a time, and we're sometimes good about putting these things together, and sometimes we're off the wall about making judgments about people." She cites two complicated examples: Shortly after the 2003 invasion of Iraq, U.S. troops encountered angry protestors at a Muslim holy site, and in the background of a televised broadcast, their captain could be heard ordering his troops to smile—for the crowd, not the cameras. He apparently intended it as gesture of good intentions. But it's not at all clear that the angry protestors would take it that way. In the notorious New York City subway vigilante case, teenagers smiled as they approached Bernard Goetz demanding money. Goetz interpreted their smiles as a predatory taunt, and he pulled out his handgun and shot. "There is no way in that circumstance that smiling is a sign of bonding," says LaFrance.

If Dan Hill quotes Oscar Wilde, she prefers Herman Melville, who said, "A smile is the chosen vehicle for all ambiguity."

Ekman dismisses his critics as diehards unwilling to acknowledge the importance of biology in shaping human behavior. He characterizes national differences, such as the British stiff upper lip or Japanese emotional concealment, as "cultural display rules." They may modulate the time, place, or degree of a display, but not the universal biological facial expressions.

He also argues that the best answer to the complexities of the human face lies not in computers but in the human mind—particularly in a handful of minds with a special gift for reading faces. Over the years, Ekman has identified about thirty such savants, people who consistently score 90 percent or better on his one-hour test of their ability to detect lies. He calls this group the Diogenes Project, after the Greek philosopher who peered into the faces of fellow Athenians by lantern light in search of one honest man, and he puts enormous faith in them. Most are in law enforcement, where Ekman does much of his consulting.

But even the Diogenes types are not "walking lie detectors," says one member of the group, Los Angeles County Sheriff's Department Sergeant Robert Harms. They're simply people who have developed the habit of listening harder and watching more closely, because their work depends on it. "It isn't magic, it isn't voodoo," says Harms. "It's just one other tool that we have in our bag of tricks."

If Harms sounds cautious, it's because he's leery of inducing false confidence—particularly among people desperate for a simple way to detect lies. The "Pinocchio syndrome," he says, encourages people to hope there might be one telltale facial expression as obvious as Pinocchio's nose for deceit, or danger. An article about Ekman's work in the *New Yorker,* for instance, left readers thinking that it was reasonable for a cop with the Diogenes gift to have shot and killed an approaching assailant based on "a hunch, a sense of the situation and the man's behavior and what he glimpsed inside the man's coat and on the man's face."

But being right nine times out of ten on the Diogenes test is hardly shoot-to-kill accuracy. That cop, in fact, was Robert Harms, who got out of the car afterward and held the assailant in his arms as he died. Harms says that what he actually glimpsed, long enough to read the brand name, was a can of hair spray in one of the assailant's hands, and the thumb of his other hand on the switch of a cigarette lighter. It was a weapon, a makeshift flamethrower capable of torching Harms and his partner. "It wasn't something I was reading in his face, or any kind of cues," Harms says. "It was the totality of the situation."

Most of the time, even Diogenes types need to sit down and talk to get a sense of what's really going on behind a suspect's facial expressions. "You're not looking for lies," says J. J. Newberry, a retired federal Bureau of Alcohol, Tobacco, and Firearms agent and one of the most consistently accurate of the Diogenes group. "You're looking for hot spots," topics that produce a flash of emotion on the face or a shift in body posture. In one arson case, Newberry noticed the torch man's lips press together in anger at the idea that he'd be taking the fall for the business owner who'd hired him. It was no more than a microexpression. But it was enough to make Newberry harp on the point until he got the torch man to blurt out the critical information. The businessman ended up with an eight-year jail sentence. The torch man served a year.

What careful interviewers look for, says Newberry, are discrepancies, places where words, facial expressions, and body language do not jibe. In his book *Telling Lies,* Ekman recounts one such moment in testimony by Vice Admiral John Poindexter during the congressional investigation into the Iran-contra scandal. Poindexter was President Ronald Reagan's national security adviser at the time, and later served under President George W. Bush as director of the Pentagon's Total Information Awareness program. During testimony he was generally unflappable.

But questions about a lunch with the CIA director produced "two very fast micro facial expressions of anger, raised voice pitch, four swallows, and many speech pauses and speech repetitions." Instead of trying to find out why the topic bothered him, Poindexter's oblivious interrogators moved on.

But if finding the truth is difficult under the bright lights of a congressional hearing, how could all of this—the weighing of words, facial expressions, and body language—ever work in a fast-paced business or a crowded subway car? Or at an airport security point with hundreds of travelers slogging past every hour?

Probably the most useful lesson we can draw from Ekman's work is simply for ordinary people to pay attention. The knack for reading faces isn't the exclusive property of Diogenes types, or members of Congress, or government counterintelligence agencies. It's an ability latent within us all—never lost, merely forgotten. With a little conscious effort, we can get it back again, though it may entail that most dreaded of interactions: making small talk with strangers in public places. It is a daunting thought. But no government security program, nor any computer network ever likely to be conceived, could possibly match the effectiveness of millions of ordinary people simply making conversation and looking one another in the face.

A few years ago, for instance, a U.S. Customs agent named Diana Dean bent down to question a motorist on a ferry in Washington State. She asked four simple questions. His answers didn't make sense, and Dean noticed that he was also acting, as she put it, "hinky." She had her inspectors open the trunk, and they found nothing. Then they looked in the spare tire well. It contained packages of a powdery substance and a big jar of a honey-colored liquid.

The driver jumped out of his car and ran, but the inspectors caught him and locked him in the back of a patrol car. As they turned their attention back to the trunk, Dean noticed that the suspect was now lying down in the patrol car, and every once in a while he would peer up over the door rim with wide eyes. One of the other agents was holding up the jar and whirling it around. They guessed it was some kind of drug.

In fact, it was a type of nitroglycerine, which the suspect was planning to use to blow up Los Angeles International Airport. And Dean, who professes no special expertise about faces, realized afterward that the wide-eyed expression she had seen on the suspect in the patrol car was fear. He was lying down in the backseat to protect himself, because he thought they were all about to be blown to smithereens.

Most of us are, of course, unlikely ever to discover a terrorist or save countless lives, as Dean surely did. But the moral of her story is clear: Pay attention to faces, and even when you think the deal is done and the bad guys are safely locked away, keep paying attention to faces. They are our best window into other people's inner beast.

PARTY FACE

When your boss comes at you with anger visible on his face, his expression derives directly from what primatologists call the "compressed-lips face" in chimpanzees and the "tense-mouth face" in macaques and baboons. Point this out and you may observe the expression known in monkeys and apes as the "scream face." The faces of other primates are often richly expressive in ways that remind us of ourselves. For instance, the guenon, a brightly colored leaf-eating monkey in West Africa, has what one biologist calls "smashing, artistic facial makeup." Like many other species of monkey and ape, guenons also lip-smack. It's a ritualized imitation of the grooming behavior in which one monkey reaches in with his lips to pick insects or debris from another monkey's hair. Lip smacking from a distance is a way to say, "I want to groom you, I want to be your friend, I'm not a threat." We say exactly the same thing, and in almost the same way, when we greet friends and fellow workers with an air kiss.

11

FACIAL PREDESTINATION

How the Shape of Your Face Can Make or Break Your Career

Everyone sees what you appear to be, few really know what you are.

—NICCOLÒ MACHIAVELLI

Physiognomy is destiny.

Physiognomy is also nonsense.

This is one of the weirder contradictions to arise from looking at ourselves as animals. From ancient Greece to nineteenth-century England, physiognomy was a high science, as systematic as it was illogical: Practitioners held that they could read a person's character not in fleeting facial expressions but in the flesh and bone of permanent facial features. Aristotle believed, for instance, that the size of a man's eyes revealed his temperament. Likewise, the captain on HMS *Beagle* almost missed his chance at immortality because he thought Charles Darwin's nose indicated a lazy character, unsuited for a round-the-world voyage.

We know better now, just as we know not to put much faith in astrology. And yet all of us still unconsciously practice the bogus science of judging character on the basis of facial features. We hire (and sometimes fire) people because of physiognomic prejudices of which we are scarcely even conscious. It's why your company may be run by a person whose chief qualification is a pronounced resemblance to a Ralph Lauren model. It's why your rival with the more mature face advances steadily through line management, whereas you, with your chubby cheeks, have somehow gotten shunted off into human resources.

Our faces are our destiny, and evolution is largely to blame. We are biologically prepared to connect certain facial expressions with specific emotions. And then we overgeneralize, treating permanent facial features as the mark of the emotional style they accidentally resemble. Because the lips swell during sexual excitement, for instance, we seem to regard bee-stung lips as a sign of sexual readiness. So we act as if full-lipped people constantly flutter on the cusp of passion, whereas thin-lipped people wear out their mattresses only by leaning on them during bedtime prayers.

Similarly, we have evolved as mammals to coo over a baby face; it's how nature tricks us into taking care of our offspring. When baby-faced features carry over into adulthood, a phenomenon biologists call neoteny, our innate response carries over, too. We are likely to be more open and trusting with such a person. So someone with a baby face may actually be better suited to work in human resources than, say, someone with a severe brow.

And this is the really insidious thing: Because people respond to our faces in predictable ways, it may gradually shape how we behave and who we become. So it can seem as if physiognomy really works.

WARRIOR APES

For instance, one study categorized graduates in the West Point class of 1950 on a scale from more dominant-looking (typically meaning prominent brow and strong jaw) to more submissive (typically with large eyes, high, thin eyebrows, and round face). It didn't make much difference as they progressed through the middle ranks, where advancement was often decided by remote, impersonal promotion boards. But facial type played a significant part in determining who ultimately became a general. The top rank overwhelmingly went to individuals with dominant-looking faces.

One possible explanation, researchers Allan Mazur and Ulrich Mueller suggest, is that these faces resembled human and nonhuman primates "preparing for a fight: thin lips, withdrawn corners of the mouth, lowered eyebrows with partially closed eyes (in order to protect them against injuries), withdrawn ears, making them appear smaller." Thus the highest ranks "perhaps go preferentially to those who present themselves as tough warriors." People who make promotions apparently treat facial dominance as a reliable indicator of "the capability to actually dominate other people, even if this ability in the social context must be applied only in a subtle and purely symbolic way." And this irrational assumption may well be right: Subordinates seem to prefer their leaders to look like General Patton, not Howdy Doody.

Having a baby face can have its advantages, too, though mostly on a different order of achievement. A Boston study tracking more than five hundred cases in small-claims court found that defendants with big eyes and chubby cheeks generally got off. Judgments in cases of intentional wrongdoing went against mature-faced defendants 92 percent of the time but against baby-faced defendants only 45 percent of the time. The perception, according to Brandeis University psychologist Leslie Zebrowitz, is that "baby-faced people are too honest and naive to have a high probability of committing a premeditated offense." So they may get away with murder, or at least petty theft. If you are the aggrieved plaintiff in such a case, you may be tempted to use the proper scientific terminology for addressing your rival, which is "Look, you neotenous little creep . . ."

But don't let the judge hear it. In fact, you should probably refrain from this whole line of thinking, because, intuition to the contrary, making judgments about people based on accidental facial features is dumb. We know it is dumb because so many organizations opt for the General Patton look-alike only to find out later that he is Howdy Doody on the inside. He does not know what a wing nut is, even when he is looking in the mirror. He could not lead a starving dog to his dinner bowl. But because he looks good on television, the so-called face guy, a walking, talking Potemkin village of a human being, abounds in the upper ranks of important companies. Zebrowitz ascribes this sort of thing to the "facial fit principle." It's the tendency to steer people into jobs for which their faces seem to suit them, and until the late 1940s, companies did it systematically, using the Merton system of face reading to detect the stereotypical markings of executive style.

Even now, some companies hire proponents of facial predestination to help them make such specious judgments. Mac Fulfer, a former Texas trial lawyer, says he has sold his face-reading services to such companies as Sprint, Sun Microsystems, Bell Helicopter, Quaker Oats, and New York Life. He started out using the lines of the face as a shorthand guide to the personalities of potential jurors and now teaches companies that they can apply the same technique to hiring new employees.

His clients pay for his advice because Fulfer seems at first to make sense. The lines in your face did not get there by accident, he says. They come from the way you move your facial muscles, "the held-on, habitual patterns of thought or feeling." He also avoids absolutes: "I don't think you can use anything, the Myers-Briggs test, or face-reading, or handwriting analysis, or anything else that you can really nail somebody and say, 'This is you.' Because people are too complex. I'm not trying to nail people down. I'm just trying to say how you've been using those muscles."

When he gets rolling, it's a bit like listening to a quick, and surprisingly insightful, fortune-teller. The high eyebrows in a photo of former Hewlett-Packard boss Carly Fiorina indicate that she is "a strategist, she takes information in very carefully, then comes up with a plan. She doesn't like to be stampeded." Her close-set cheeks mean "tremendous energy for short bursts." Then Fulfer adds that her inner ear ridges reveal that she is capable of "great pattern recognition," and we are back in the land of Darwin's lazy nose.

Fulfer once sat in on a dinner with a prospective vice president at a veterinary medicine company, and the candidate got the job in part because Fulfer said he had "managerial eyebrows." (When auto industry executive Robert Lutz was rising through the ranks, a GM executive likewise once admonished him for having a Marine-style flattop instead of traditional "executive hair." The company evidently followed the *Dilbert Principle* theory of how people become managers: "Well, he can't write code, he can't design a network, and he doesn't have any sales skill. But he has very good hair.")

Fulfer has also advised clients that people with continuous eyebrows (the unibrow, to you and me) are "powerful thinkers whose minds are always at work. Let them tell you some of their ideas." And: "We see jowls here, a person who's comfortable exercising power and authority, incredible stamina. Once they lock on, they're bulldogs." And finally, this irresistible physiognomic nugget: "One of the things I see is, look how large his nostrils are. So he comes from a space emotionally of abundance."

One can almost hear the executive search committee reporting to the board of directors at Blodget.com: "This is our top candidate for CEO, and we recommend a $16 million stock-option bonus package on account of his extraordinarily capacious nasopharyngeal passages." It would explain a great deal about modern corporate history.

KEY STIMULI

Hiring a new employee in this fashion is about as prudent as the way families choose a new dog at the local pet shop: Drawn in by the ineluctable siren song of all things baby-faced, you fall for the puppy with the big brown eyes and the smooshed-up nose. Two weeks later, after the kids have already bonded, it dawns on you that the companion you have brought into your home for the next ten years is a mean, yapping, ankle-biting son of a bitch. Congratulations! You have just become the latest victim of a biological scam known among ethologists—students of animal behavior—as "key stimuli," simple visual or vocal signals that elicit an automatic, unthinking response. In the case of the baby face, the responses include increased caretaking behavior and reduced aggression among adults who probably ought to know better. Key stimuli are commonplace in the animal world, and they are startling for their effectiveness. In her 1997 book *Reading Faces,* Zebrowitz cites a poignant example: A turkey chick's call elicits nurturing behavior from its mom, and this is an innate response, not something the mom thinks about too much. If she is deaf and cannot hear the key stimulus, she is liable to kill her own offspring. On the other hand, if she can hear the chick's call (or a researcher's tape recording of it) from a stuffed polecat, the turkey mom will rear the polecat—a natural predator of turkeys—like her own precious little ball of down.

Primates are somewhat more thoughtful than turkeys but also respond to key stimuli. In baboons, for instance, infants wear a black coat until about the age of twelve weeks, and then turn to the mousy brown of an adult. The mom protects and nurtures the baby as long as it remains black, but becomes indifferent when it takes on adult coloration. Monkeys and apes also use facial features, particularly the round, wide-eyed baby face, as key stimuli. So do dogs, rabbits, birds, and many other species where infants need to elicit adult care.

THE ATTRACTIVENESS HALO

In the workplace, it is generally prudent not to act like a turkey. So the lesson to learn about key stimuli is that we should pay *less* attention, not more, to permanent facial features. And the way to do that is to be aware of our biases, our evolutionary predispositions, so we can get beyond an individual's physical appearance and look clearly at his or her behaviors.

For instance, everybody is a sucker for what psychologists call "the attractiveness halo." We have this idea in our heads—in our genes, really (yeah, okay, and in our jeans, too)—that attractive people are actually better people. This is partly what compels us to pay slavish attention to endless identical television reports about how Charlie Sheen or Drew Barrymore is learning to take life one day at a time (down from three-day jags, though they still cannot remember any of them).

In one study, researchers gave some men a photograph of an attractive woman. Other men got photos of an unattractive woman. Then they talked to a woman on the telephone. The men who were operating on the assumption that she was pretty treated her as if she were also warm, intelligent, and outgoing. (Observers asked to rate the conversation afterward said the woman responded more warmly as a result.) The men who thought they were stuck on the phone with a skank were, oh, less generous.

This isn't very nice. But before you start fulminating about what miserable bastards men are, you should know that babies do it, too, even as you are goo-gooing over their adorable chubby cheeks: Infants as young as two days of age respond more negatively to unattractive faces.

This built-in prejudice is entirely natural. All species evolve through sexual selection (individuals outcompeting their rivals for the affections of the opposite sex) as well as through natural selection (managing not to get killed). Some of the things that make people socially attractive and sexually selected are cultural and may change from year to year, like nose rings or Armani suits. But others are biological and permanent.

We have evolved to regard symmetrical faces and "average" faces, for instance, as more pleasing. This may be because crooked faces suggest a hazardous level of inbreeding. In any case, people with attractive faces get more smiles from babies and more dates with adults than people who are, say, wall-eyed.

They also tend to get better jobs. Numerous studies demonstrate that being ugly is hazardous for your career.

HIRING THE UNATTRACTIVE

In some cases, appearance is a bona fide job qualification. "If you're selling cosmetics, maybe you want to have a beautiful woman," Zebrowitz concedes, "because who's going to want to buy cosmetics from an ugly woman?" It is also entirely legal to discriminate against people because you think they have been hit with the ugly stick, so long as your idea of ugly (a word derived from the Old Norse *uggligr,* "causing fear") is not based on race, age, religion, or gender. But this is a tricky line to walk.

The clothing retailer Abercrombie & Fitch, for example, recently paid $40 million and agreed to hire diversity recruiters to settle a class-action lawsuit alleging that it staffed its stores with hip, attractive young people who also happened disproportionately to be white.

More important, it can be wasteful to discriminate, even if somebody's face makes you wince, because physical attractiveness is irrelevant for many jobs. Given a modicum of respect, a computer programmer with pebble-dash acne scars can do his work at least as well as a handsome one, and may well be more motivated to stick with the job.

And yet the prejudice against ugly people runs deep. "I believe it is true that attractive people are actually better people," one executive in the wireless telecommunications industry advised me by e-mail. "They are not born better, but they become better socialized by virtue of the way they are treated as they grow up. Ugly people develop complexes due to the humiliation and abuse they endure over the years. It's like dogs: beat one and he turns mean; love one and he loves everyone. I generally believe that employers need to be careful about hiring the unattractive."

This put the case so nakedly that I was preparing something between a reply and a rebuke when another e-mail arrived, from an internal auditor at a major technology company: "I am no Ralph Lauren model. I do not have a strong chin. Neither do I have a prominent brow or a strong jaw. I am short and a female. I am loud." It was impossible to mistake the rancor welling up from a lifetime of being overlooked or abused on account of key stimuli: height, attractiveness (or the lack of it), and gender. "For 30 years, I speak in a loud manner," she wrote. "Trust me, everyone can hear me. I demand attention. If I don't get your attention, everyone will know, because I can be heard five aisles away."

She added that she was the lead plaintiff in a class-action lawsuit against her employer's pension plan. In fact, a court had decided the case in her favor, with a potential cost to her employer, pending appeal, of billions of dollars to undo some overly clever pension plan accounting from the 1990s. (I asked around, and her story checked out.) You could read this in either of two ways: It was a powerful argument against hiring unattractive people, or it was a costly reminder that clever accountants and executives need to be careful about heaping humiliation and abuse on the unattractive people they inevitably hire. I prefer the latter reading.

It may be difficult in practice, but as policy it makes sense to try to avoid biological traps such as the appeal of baby faces or the attractiveness halo. But how? Zebrowitz advises employers to put off the face-to-face meeting and learn as much as possible about a potential hire by other means—the application, letters of recommendation, e-mail notes or instant messaging, and telephone conversations. What works for romance can also work for business relationships: People now often go through the preliminary stages of dating by electronic means, as a way to determine how the other person really thinks, without the fleshy distractions of sex appeal. Establishing a rapport beforehand can be a way to take the dazzle, or the sting, out of misleading first impressions.

Reminding yourself of potential bias can also help, says Zebrowitz: "Am I thinking this person can't handle a management job because she really can't? Or am I biased by her appearance?" The bias need not be explicit to become a factor in hiring. In one study from the 1970s, white interviewers displayed nervousness and signs of greater social distance with black applicants, with the result that blacks fared poorly. When the researchers trained the interviewers, a little perversely, to repeat this nervous, distant interviewing style with all applicants, whites also performed badly.

Zebrowitz suggests that this sort of unconscious distortion of the hiring process may extend to matters well beyond race. Interviewers may be more open with baby-faced candidates, for example, or toss softball questions to someone who "looks like a leader."

In one experiment, Anke von Rennenkampff, a graduate student working with Zebrowitz, gave study subjects a list of eighteen questions they could ask candidates for a leadership position, nine positive ("What was your greatest sense of achievement in college?") and nine negative ("What was your worst failure in college?") It turned out that the more masculine-looking the candidate's photograph, the more positive the questions. "Participants didn't want to find out more about the applicant," says von Rennenkampff. "They wanted to find out that their first impression was correct. They wanted to confirm their opinion, not get objective information."

FIDDLING WITH FACIAL PREDESTINATION

What if you are the job seeker caught in the baby-face trap, or otherwise a victim of your facial features? What if you are—let's be frank—stump ugly? The temptation is to change yourself, and this may sometimes make sense. "A man asked me, 'Why do people always think I'm weak?' " says psychiatrist Paul Ekman. "So I looked at his face, and his eyebrows grew in such a way that they appeared to go up at the inner corners." That is, they accidentally echoed the facial expression for sadness and uncertainty. "I told him, 'Pluck your eyebrows,' and it worked." Men can disguise a weak chin with a beard, Zebrowitz adds, and women can darken their eyebrows with cosmetics for a more mature appearance.

The tradition of putting on your best face, and also embellishing it a little, is entirely honorable. Sir Isaac Newton circulated portraits of himself with piercing eyes and a broad, intellectual brow, undeterred by a friend's remark that in real life Newton's face showed "nothing of that penetrating sagacity." Soap opera actors and newspaper columnists likewise routinely present themselves in head shots that are twenty years out of date, and why not? It is frankly annoying to think that physiognomy is destiny and that we should be victimized by the evolutionary foibles, the unconscious, inherited biases, of our fellow human beings.

The face you choose to put on might also reasonably vary depending on the type of job you are seeking. People seem, for instance, to prefer masculine features in a leader. This may be a genetic predisposition, rooted in our biology as a male-dominated species. Or it may just be a bad habit from being bossed around for the past ten thousand years by what Carly Fiorina, the former CEO of Hewlett-Packard, once called men "with twenty-inch necks and pea-sized brains."

Either way, the taste for masculine-looking leaders holds true *even when all the potential candidates are women.* Researcher Anke von Rennenkampff asked 120 people to imagine that they were hiring someone for a leadership position. She gave them applications that were identical except for the accompanying photograph. One candidate was a woman with a biologically masculine face (high, square forehead, bigger chin, thin lips, eyebrow ridge more protruding). Another candidate was a woman with a feminine face (rounder cheeks, a smaller nose, big, round eyes). Von Rennenkampff used two photographs of each woman, one in a masculine style of dress (black turtleneck, no makeup, hair pulled back), the other in a feminine style (loose hair, lipstick, a necklace, a pullover with a scoop neckline). Predictably, the candidate who looked the most masculine, in both biology and dress, got the leadership job. The one who looked doubly female fared worst. It was a clear case of the facial-fit principle at work. When von Rennenkampff asked participants to choose candidates for a human resources job requiring communication skills and sensitivity to others, they reversed themselves, as the facial-fit principle would also predict, and gave top rank to the doubly feminine candidate.

The one promising aspect of the study was that if the feminine candidate wanted the leadership job, she could improve her standing from number four to number two simply by dressing in a more masculine style. With a little tinkering, in other words, it's possible to mitigate the effects of facial predestination. This is a lesson von Rennenkampff herself has plainly taken to heart. She is a tall, slender young woman with big brown eyes and smooth skin. But she wears her hair pulled back, not much makeup, small stud earrings, a long-sleeved taupe blouse under a dark sleeveless sweater, and dark slacks.

It's also possible to escape a facial stereotype by changing your behavior, or even just your facial expression. The stereotypical effect of a facial feature typically lasts only as long as the face is in repose. Once the face takes on an expression, everything changes.

For example, I have a severe brow, which often gets mistaken for anger. Women sometimes hesitate to get on an elevator alone with me. Fortunately, I also have a big, appealing smile, and I like to think the contrast is disarming. (Imagine a radiant dawn breaking over the garbage hills of the New Jersey Meadowlands, my native haunt.)

Women still hesitate to get on an elevator alone with me. But Zebrowitz argues that these "contrast effects" can make certain behaviors more impressive simply because they are so unexpected. A woman may seem like a better leader, for instance, because her assertive manner is so contrary to her baby-faced appearance. A man who looks like General Ursus in *Planet of the Apes* may endear himself with a small, unanticipated act of conciliation. (Imagine what it might be like, for instance, if something nice made Donald Rumsfeld smile.)

While cosmetic surgery may be tempting, a healthier approach is to stick with the face your mama gave you and work shrewdly with the predictable ways people respond to it.

▪ As jowls and crow's-feet creep over your face, you can take them as a public announcement that you have passed your sell-by date. Or you can use them as tools for eliciting greater respect. Older lawyers often employ what one overbilled client describes as a "froglike folding down of the mouth's corners simultaneously with an elevating of the pursed lips and thrusting of them forward, as a way of looking judicious. Or contemplative. Or doubtful." Truth may be beauty, and beauty truth. But in the legal profession, billable hours are better than either, and looking jowly and wise is a good way to earn them.

■ If you have a baby face, you may be better at getting people to trust in you, often a useful advantage. That was the effect Kenneth Lay had at Enron, with his soft eyes and self-deprecating manner. On the other hand, the contrast effect can sometimes lend a sinister or treacherous air to baby-faced individuals who get caught in some unseemly behavior. That's why the baby-faced killer is such a compelling figure in crime reporting.

■ If you have a shrewder, more mature face, you may want to enlist a baby-faced colleague, or bring in a warm, sympathetic human resources type, to inform the staff that you have just outsourced their jobs to Chongqing. ("Could I interest you in a job transfer and a 99 percent pay cut?") That's because baby-faced individuals are less likely to elicit sociopathic rage. This is a strategy deeply rooted in primate behavior; it's also familiar to any child who has ever enlisted a cute younger sibling to tell Mommy about having smashed her favorite vase.

It is, of course, also possible to mitigate the effects of facial stereotyping with the help of a surgeon's knife. We live with the myth that almost any sow's ear can be turned into a silk purse—preferably via cosmetic surgery on a reality television show devoted to extreme makeovers. Fulfer cites the case of a Texas bus company ticket clerk whose career was going nowhere because of a receding chin. Then she got a chin implant and eventually rose to become president of the company.

The problem is that, at least outside the bus industry, people are extraordinarily skillful at detecting fakes. So your rival down the hall will probably make a point of naming the surgeon who bobbed your nose. Your enemies will whisper at the water cooler that your upper lip "is hideous. It must be Gortex. It doesn't even move." Worse, they will look at your collagen-inflated lips, as British tabloids have done with the unfortunate actress Leslie Ash, and label you "trout pout." Your breast enhancement may pass for the real thing but elicit innate responses of the sort that lead to the bedroom, not the boardroom. Your Botox injections may ease the knotting in your brow but also leave you looking paralyzed with indifference while your coworkers arch their eyebrows in surprise and delight at the boss's every witty utterance.

The tantalizing promise of cosmetic surgery is that you may perhaps escape the stupidity of facial stereotypes. The danger is that you could end up, like Michael Jackson, looking almost as human as a ventriloquist's dummy.

Your call.

Infant chimps have white tail tufts until about three years of age, and adult males indulge them and tolerate their pestering as long as the baby trait persists. Under the protection of this key stimulus, an infant chimp gets away with it even when he tries to push Big Daddy off Mama during copulation. This is a case where extrapolating from animals to humans is particularly hazardous. So just to be perfectly clear: Even the most baby-faced employee should not attempt this behavior on discovering the secretary and the boss *in flagrante delicto.* But you should expect a nice promotion (to a leadership position) in the new year.

MONKEY SEE . . .
The Power of Imitation

We worship not the Graces, nor the Parcae, but Fashion. . . . The head monkey at Paris puts on a traveler's cap, and all the monkeys in America do the same.

—HENRY DAVID THOREAU

"I don't remember what the women wore," Lou Gerstner noted after his first meeting with top management upon becoming the chief executive at IBM, "but it was very obvious that all the men in the room were wearing white shirts, except me. Mine was blue, a major departure for an IBM executive!" When the same group met again a few weeks later, Gerstner wore a white shirt as a way of fitting into the company culture.

Unfortunately, all his subordinates showed up in colors.

It's always pleasant, but a little too easy, to make fun of pinstriped business types for their conformity. The truth is that almost all social animals imitate one another. It's not just monkey see, monkey do. "Chickens who have eaten their fill begin to eat again when they are placed with a hungry chicken who is pecking voraciously at a pile of grain," writes Elaine Hatfield, a University of Hawaii psychologist and coauthor of *Emotional Contagion.* "Ants work harder when paired with other worker ants."

And humans? Despite our vaunted individualism, we are the most imitative animals on earth. We mimic the jump shot of the ballplayer on the television screen or the crestfallen face on the tsunami victim who has just lost her only child. We crack up with the encouragement of a laugh track.

It's one of our most basic and underrated biological drives: We want to be like other people, or at least like people we identify as kindred spirits. It's why corporate lawyers dress like other corporate lawyers, and anarchists like other anarchists. It's why reporters are mostly rumpled slobs who deeply miss the sartorial refinement of ashes down the shirtfront. We actually go out of our way to create circumstances in which we can imitate other people. How else to explain an otherwise deviant pastime such as line dancing at the company picnic? How else to understand the peculiar pleasure of factory workers in Beijing doing tai chi together before the start of the morning shift?

The power of imitation is immense and highly adaptive. Watching the people around us teaches us how to do the job right, and also how to behave as a social species, to fit in with the group so we don't get left behind. During much of our history as tribal animals, failing to conform with the local culture was a quick way to die. ("Oh, man, that freak won't paint his face blue! Let's toss him to the wolves.") So natural selection ensured that conformity and the imitative urge got embedded in our genetic heritage. Special cells in the brain, called mirror neurons, help us unconsciously mimic the people we meet and in the process share their emotions. We hardly even think about the practical purpose anymore; imitation simply feels good. Doing things in synchrony with other people produces a deeply gratifying hum in some obscure corner of our minds.

BORN TO IMITATE

Infants begin mimicking facial expressions within one hour after birth, and we go on imitating words, faces, body language, styles of dress, social fads, and fashions until we die. Just in terms of vocalizations, Hatfield lists a "staggering" number of characteristics people in a conversation do not merely mimic, but mimic simultaneously, including accent, speech rate, vocal intensity, vocal frequency, pauses, and quickness to respond. If the dominant person in a conversation has a rat-a-tat-tat speaking style, the others pick up the pace, too. If he talks loudly, everybody else also tends to bellow. These behaviors show up not just in controlled laboratory experiments but also in "tightly structured job interviews, presidential news conferences, astronaut-to-ground communications," and ordinary office chitchat.

In *The Right Stuff,* for instance, Tom Wolfe describes how test pilot Chuck Yeager transformed the way an entire generation of airline pilots spoke because everyone wanted to emulate his coolness under pressure. That voice "drifted down from on high, from over the high desert of California, down, down, down, from the upper reaches of the Brotherhood into all phases of American aviation . . . Military pilots and then, soon, airline pilots, pilots from Maine and Massachusetts and the Dakotas and Oregon and everywhere else, began to talk in that poker-hollow West Virginia drawl, or as close to it as they could bend their native accents. It was the drawl of the most righteous of all the possessors of the right stuff: Chuck Yeager."

Down, down, down on terra firma, the conversational imitation also takes baser forms. One executive coins an utterly unrighteous verb such as *incentivize,* and all the other executives are soon nattering about how they can "incent" people to do more work, usually without paying them an extra dime, but earning handsome bonuses for themselves. (Perhaps the word they are looking for is *incense*?)

We mimic each other even when we are just trying to get home after work, slowing down on the highway for some disturbance in the traffic flow—for instance, a rush-hour merge from a busy on-ramp—and thus creating a kind of phantom traffic bottleneck that persists, through imitation, long after the original cause has faded away. Hence the maddening experience of arriving at long last at the end of the congested area only to find nothing there. No fender-bender. No ambulance. Not one damned reason to cause a traffic jam. We speed up, muttering about the stupid bleeping drivers ahead of us. Five minutes later, other drivers arrive at the same dismaying, imitation-induced time warp and entertain identical bad thoughts about us.

We match each other's stride so closely that armies traditionally learned to break step when crossing a bridge, lest they make it vibrate to the point of collapse. We mimic each other even at the biochemical level: Women who work close together unconsciously engage in a kind of pheromonal conversation, gradually drawing one another's menstrual cycles into synchrony.

There are plenty of theories about these herd phenomena, and not all of them speak of sisterly love or the need to resonate with feelings of oneness. It may be, for instance, that we enjoy being with other people, and being like them, partly for safety in numbers. In a paper titled "Geometry for the Selfish Herd," the biologist William Hamilton argued that animals often form tight flocks, schools, packs, and herds because each individual is trying to keep at least one other individual between himself and the wolf at the door. (Unfortunately, every defensive tactic elicits a predatory countertactic. Sea bass, for instance, race straight at the center of a school of prey fish, split the school in two, then strike at the tail of one of the new groups. Fish formerly ensconced in the center of the school are thus among the first to die.)

Doing exactly what everybody else is doing, no matter how dumb it may look to outsiders, is often a way to avoid danger. For instance, Arctic seabirds called guillemots nest on steep cliff faces. Fledglings leap off the cliff and into the sea en masse. Why? If a couple of fledglings leap by themselves, they will almost certainly get picked off by predatory glaucous gulls. If everybody leaps together, some of them will still get eaten, but the gulls will be too distracted or too full to bother with the rest. One biologist has dubbed this the "feed them your neighbors" strategy.

Defensive herd impulses of this sort are deeply ingrained in our biology. This is one reason it's so difficult to get individuals or companies to try something new, or to stop doing something familiar that always worked in the past. We have this subconscious feeling that if we aren't doing what everybody else is doing, then we are probably about to die.

And it's sometimes true. For instance, mirroring the facial expressions of people around us can be a survival mechanism. Let's say you're standing around listening to a coworker gripe about how your boss, Thimblebrain, goes out of his way to make her feel stupid. You happen to look up and spot Thimblebrain bearing down on you. An expression of alarm flashes across your face. (It takes just two hundred milliseconds, one-fifth of a second, from threat to facial expression, as you may recall from an earlier chapter.) The identical expression leaps from your face to the faces of the people around you, causing them to feel fear, too.

The conversation dies away just in time, and all this happens before anybody can say, "It's him."

Indeed, because the conscious mind needs five hundred milliseconds to recognize a threat, the whole defensive exchange could theoretically occur, and enable all of you to keep your jobs, before anybody actually has a clue about what's frightening them. Nor is this subconscious form of communication limited to fear. Try wrinkling your nose with disgust during lunch in the company cafeteria, for instance, and notice how fast everybody else stops eating, forks quivering in midair. Rapid emotional display and response of this sort—also known as emotional contagion—is part of what makes us such a success as a social species.

It does not, however, by any means make us unique. In one study, a rhesus macaque monkey learned to recognize a visual signal as a warning that he was about to receive an electric shock. The monkey could avoid the shock by pressing a lever. A second monkey in a separate cage couldn't see the warning signal. He could, however, see a black-and-white video image of the first monkey's face. When the first monkey's eyes widened in fear, the second monkey leaped to press his lever, thus also avoiding an electric shock. Curiously, this survival mechanism wasn't evident in monkeys raised in isolation. Primates like us seem to need a normal social upbringing to learn how innate facial expressions can save our necks.

SEX AND THE CROWDED PARKING LOT

For animals and humans alike, there are also more positive reasons why we often act like birds in a flock. Imitating others is frequently a shortcut to success. Think about the familiar rule of all road warriors: Never stop at a diner with an empty parking lot. The logic of choosing someplace crowded isn't that we want to wait for a table. We merely treat parked cars, the hubbub of other people, and even the inconvenience of a line as evidence that the food is worth waiting for. Or at least that nobody has died there lately of *E. coli* poisoning. Animals do the same thing. Starlings, for instance, like to probe for food where other birds are already probing successfully. Rats seek out the sweet morsels they've sniffed on the breath of their nestmates.

The tendency to imitate even affects how we choose a mate. Guppies, ruffed grouse, and single women in Manhattan all prefer to bypass the lonely guy for the guy who's already got a hot date.

It's the diner parking lot theory in a different form: Females treat a male's ability to attract other females as evidence that he may actually possess attractive qualities. A New York entrepreneur recently started a kind of zipped-up escort service on this premise: Men pay $50 an hour to hang out with an attractive, personable woman in a public place because, as the company, Wingwomen.com, puts it, "Every guy out there knows that it is much easier to meet women when you're around other women."

We also imitate one another because it elicits comforting feelings of closeness. Two people in a friendly conversation, for instance, frequently match each other's body language down to the crossing of their ankles or the waggling of their feet. This postural echo is one way we communicate the sense of affiliation and affinity. When it happens unconsciously, it feels good for both partners, as a way of saying, "I'm with you."

Studies demonstrating this effect are generally designed so that the subject is not conscious of having been mimicked. Even so, the results suggest we like a conversational partner more if the other person has subtly mimicked us. Being mimicked also makes us more fond of people in general; we seem to bond with the entire human race.

These may sound like airy-fairy feelings of oneness, or like an expression of raving, insatiable ego. Skeptics may imagine the much-mimicked CEO dancing across the cubicle tops, strewing rose petals and reciting Whitman in a plummy vibrato: "I sing the body electric, / The armies of those I love engirth me and I engirth them . . ." (Dutiful senior vice presidents meanwhile look up in awe and mouth the words, wondering if they, too, can make their feet be happy.) But the sense of affiliation produced when people mimic one another can have highly practical effects.

For instance, imitation seems to help people synchronize their movements. One person at the table reaches for his drink, and a moment later everyone else also takes a sip. One person in the trading room stands and stretches, and his neighbors also stretch, and enjoy a short social break together. One reporter in the newsroom starts hammering away on a story, and the entire staff soon stirs to life, like worker ants, in a contagious frenzy as the deadline draws near.

Anthropologist Edward Hall describes the construction workers putting an office addition on his house. "Conversation was continuous. It never stopped. Yet the content was not highly relevant. They were talking to be talking. If the conversation lagged, the work lagged. Two or three men could work in a very small area without ever seeming to interfere with each other, and they worked very close together. Whether adobe bricks were being laid, plaster was being applied to the walls, or cement was being smoothed, the whole operation was a ballet, with the rhythm of the conversation providing the unconscious score that strengthened the group bond and kept them from interfering with each other."

A team that finds the rhythm of its work in this manner coalesces, so the parts of a project get handed on seamlessly, as if by magic. One person starts a sentence and the other person finishes it. One comes up with a new product idea, and the other nudges it in a brilliant new direction. It's like the behind-the-back pass to the trailing player who arrives just in time to catch the ball and complete the lay-up. Or like musicians who wander off to the far corners of the stage for their solos, then slowly drift back, unbidden, all of them arriving at their microphones at the same instant and in the same key for the first line of the chorus: "I'm gonna tell you how it's gonna be . . ." You get a chill down your spine and your hair stands on end, because we are built by evolution to relish these moments of perfect synchrony.

The curious thing is not just that people mimic one another but that we observe mimicry and use it to draw relatively sophisticated social judgments without ever apparently being conscious of it. In one experiment, test subjects viewed fifty one-minute clips, each showing a different man and woman talking across a table. The task was to decide how much the partners really liked each other. The test subjects said afterward that they relied on how close the partners sat and how often they smiled, nodded, gestured, or otherwise expressed their feelings. Two-thirds of them said they paid no attention whatever to whether the two people mirrored each other.

Oregon State University researcher Frank Bernieri and his team then went back to the one-minute clips and made a detailed record of seating proximity, smiling, gesturing, and mirroring behaviors (or synchrony). It turned out that the couples most often judged to like each other weren't necessarily the ones who smiled or nodded the most—and that test subjects relied on synchrony far more than anyone imagined.

Our tendency to imitate one another also has a more serious consequence: What starts skin-deep reverberates at the deepest emotional level. Putting on the facial expressions, voices, postures, and movements of the people around us causes us literally to feel what they feel.

In the 1960s, when psychologists Paul Ekman and Wallace Friesen were using their own faces to chart the muscle movements involved in different expressions, they kept finding that particular expressions caused them to experience the corresponding emotions. The smile that indicates genuine happiness (with crinkling at the corners of the eyes) made them genuinely happy. An expression of grief caused them to dwell on their own losses. Numerous studies since then have established this feedback loop as a biological fact: Emotions produce facial expressions, and facial expressions in turn produce or intensify the associated emotions.

The biological mechanisms involved may be surprisingly simple and direct. For instance, smiling literally cools the brain. The muscle movements needed to produce a smile increase the volume of air inhaled through the nose, reducing arterial temperature in the region. A slightly cooler blood temperature reaching the brain puts subjects in a better mood. Fierce executives practicing their "mean business" scowls may thus be doing precisely the wrong thing if, as they often suggest, their goal is to achieve coolheaded decision making.

THE INFECTIOUS WORKPLACE

Emotional contagion—picking up the facial expressions and the emotions of the people around us—is a pervasive force in our everyday working lives, and a particularly insidious one since it typically goes unrecognized. Health care workers spending their days with people who are sick or depressed may go home depressed themselves. Customer service workers may catch the hostile mood of unhappy customers and unwittingly pass it along to the next guy in line.

The opposite is true, too: People who spread positive feelings naturally tend to create happier workplaces. But even in emotionally neutral jobs, unpleasant emotions may be more contagious than pleasant ones, because our built-in negativity bias causes us to notice threatening behavior more quickly and respond to it more powerfully.

Psychologists Elaine Hatfield and Richard Rapson once had a client, a dentist, who was ordinarily cheerful and energetic. She managed to be nurturing without ever becoming one of those dentists who really do seem to feel their patients' pain (the ones, that is, who end up going home to drill a hole in their own heads). But one week she turned up at the psychologists' office feeling depressed. Her staffers had been squabbling, and the dentist had tried to sort out their problems by staying up late to draft a new organizational plan with carefully delineated job descriptions. Two staffers saw it and promptly gave notice.

As the psychologists questioned the dentist, they gradually learned that her hygienist was having marital problems and coming to work in tears. Her office manager was seething because she felt she couldn't delegate work to the new part-time secretary. And the secretary was feeling humiliated because the office manager didn't trust her. The problem in the office wasn't organizational, it was emotional: "As we talked, it became clear to our client that she was being blown about by this whirlwind of feelings; she was catching . . . the virus of distress."

Being aware of emotional contagion, Hatfield suggests, can give a manager a better sense of what's really driving the workplace mood and also what she can do about it. The dentist took a two-day vacation to improve her own mood. Then she went back and took her staffers out to dinner, one at a time. Rather than letting the unhappy hygienist set the mood by default, the dentist began to amplify her own upbeat emotional contagion, and everyone gradually cheered up.

The mysterious ability to be infectious in this positive sense is what makes some leaders so inspiring. For example, in his biography *A First-Class Temperament: The Emergence of Franklin Roosevelt,* historian

Geoffrey Ward quotes a contemporary's description of a visit by the incoming assistant secretary of the navy to one of the humbler vessels in his command: "Once aboard the barge [Roosevelt] showed immediately that he was at home on the water. Instead of sitting sedately in the stern sheets as might have been expected, he swarmed over the barge from stem to stern during the passage to the Navy Yard. With exclamations of delight and informed appreciation he went over every inch of the boat from the coxswain's box to the engine room. When she hit the wake of a passing craft and he was doused with spray, he just ducked and laughed and pointed out to his companions how well she rode a wave. Within a few minutes he'd won the hearts of every man of us on board, just as in the years to come he won the hearts of the crew of every ship he set foot on. . . . He demonstrated . . . the invaluable quality of contagious enthusiasm."

People who are not naturally that charismatic—pretty much all of us—can at least pause every now and then to think about how we may be unconsciously infecting the people around us with our moods. Every facial expression, every vocal nuance, is a kind of emotional sneeze. Emotional contagion occurs continuously, for instance, across the counter, from cashier to customer. In a study at thirty-nine branches of a regional bank, University of North Carolina researcher S. Douglas Pugh found that tellers who smiled more and made eye contact induced a better mood in their customers, and the customers went away feeling more satisfied.

Some managers will doubtless interpret this as justification for ordering employees to smile, like Stan the peevish restaurant manager in the movie *Office Space,* who pesters Joanna the waitress to put on more "flair." ("You do want to express yourself, don't you? Okay. Good. Great. That's all I ask.")

But this would be missing the point. Managers cause emotional contagion, too. If they want happier customers, the best thing they can do is give their employees reason to smile. (Instead, Stan gives Joanna reason to lift her middle finger in his face, in front of an entire restaurant full of customers. "All right, there's my flair," she announces. "And this is *me* expressing *my*self. There it is. I *hate* this job. I hate this *goddamn* job and I don't need it.")

One person can easily steer the mood of an entire group. Sigal Barsade, a professor at the University of Pennsylvania's Wharton School, asked small groups to decide among themselves how to allocate a limited pool of bonus money. Each individual in a group had to make the case for a candidate from his own imaginary business unit. Barsade planted a ringer in each group, an actor who always delivered the same speech on behalf of his candidate. But he delivered it in one of four different moods, depending on the group: pleasant and energetic, pleasant and low-key, hostile and low-key, and hostile and energetic. The ringer's speech came first in each group, and his mood strongly influenced the subsequent dynamic of the group.

Barsade wasn't particularly interested in which mood was more successful in getting the ringer what he wanted. But being pleasant and energetic predictably made the whole group happier and more cooperative. Barsade concedes that there is "some debate about whether being happy leads to better decision making." But she cites other studies indicating that a positive atmosphere makes people willing to think harder about an issue and engage in more active problem solving.

AVOIDING EMOTIONAL CONTAGION

Technology can sometimes serve as a tool for sidestepping the hazards of emotional contagion. At one insurance company, for instance, a division head got fired for losing market share, because he stuck by reasonable underwriting practice in the face of heavy price cutting by rivals. The new guy had a good background in the business, but he came into the job branded as the son of a friend of the CEO. He also had an off-putting sense of humor and a tendency to arrogance. The senior staff was still loyal to the old boss, and meetings quickly degenerated into shouting matches. Nobody compromised. The anger was contagious and self-perpetuating.

Then one of the executives suggested using e-mail as a way for all sides to conduct their business at arm's length. E-mail has gotten a bad reputation as an arena for "flaming," the exchange of hostile fire untempered by the usual constraints of face-to-face conversation. But in this case it served as a kind of cooling-off mechanism. Since they were no longer literally getting in one another's faces, the senior staff began to think more objectively about the questions their new boss was asking.

Subtracting the human body from the dynamics of a meeting can also be a way to work around some of the hazards of social dominance. "Even though e-mail identifies the sender, something about typing on a screen equalizes people," says one Silicon Valley technology executive. In male-dominated groups, women, and those lower in the hierarchy, tend to get heard more via e-mail. People with ambiguous names, such as Lauren or Chris, can work with a group for months without anyone ever knowing the first fact conveyed by face-to-face or telephone meeting—that is, whether they are male or female.

MIMICRY FOR MONEY

The value of putting on a positive display, like the bank teller smiling at her customer, may seem obvious. But beyond that, deliberate mimicry of customers and clients has also been a common business technique at least since body language became a popular topic in the 1960s. It's part of the lore in car sales, for instance, that mirroring the customer's body language will create rapport. It's also a sometimes maddening technique among therapists to match the visual, auditory, and kinesthetic signals put out by the patient. The patient says, "I just don't get it," for instance, and the therapist comes back with, "You're feeling confused."

Some studies suggest that conscious mirroring behavior can, in fact, be effective. It may well help the patient open up to the therapist. As car salespeople suspect, such behavior may also sometimes have instant cash value. In a 2003 study titled "Mimicry for Money," Dutch researchers found that a waitress received bigger tips when she verbally mimicked her customers' orders, repeating them word for word, than when she refrained from mimicking them. The authors speculated that "mimicry may be a powerful tool in building and maintaining positive relationships between individuals" in all contexts.

But mimicry also entails hazards. Most of the time when we are automatically matching the behaviors and emotions of the people around us, it is as natural, and as subconscious, as breathing. It isn't something that requires our vaunted cognitive powers. In fact, it seems to get awkward only when we become conscious of it—for instance, when we realize in midconversation that we have just folded our hands behind our heads in precise imitation of the boss. ("When did I become such a suck-up?")

Even worse, the suspicion that someone is mimicking us can create the uncomfortable sensation that we are being mocked or manipulated. When someone stutters, for instance, the people around him often move their lips and may even stutter themselves. It's usually innocent, an unconscious act of empathy. But when a person suddenly recognizes that he is the object of deliberate and calculated mimicry, like the customer being aped by the car salesperson, it can be a deal-killer. (Make emotional contagion your friend: Get up and walk out the door. The salesperson will get up and follow, probably with a better price.) Psychologist Elaine Hatfield suggests that deliberate mimicry can easily produce discord rather than rapport. "You could imitate three things," she says. "But while you're imitating those three things, eight thousand other things are going on. So the rhythms are awkward. It's not going to be a ballet, but a robot pretending to be a human being."

Likewise, mimicking the boss may seem like a good idea. But only up to a point. Desmond Morris warns against the hazards of too much upward-mimicry: "A subordinate can, if he wishes, unnerve a dominant individual by copying his body actions. Instead of sitting on the edge of his chair or leaning eagerly forwards, he can sprawl out his legs and recline his body in imitation of the high status posture he sees before him. Even if verbally he is politeness itself, such a course of action will have a powerful impact, and the experiment is best reserved for moments immediately prior to tendering one's resignation."

Hollywood producer Brian Grazer, for instance, wears his hair in a distinctive sticky-uppy style. One day an employee showed up for work wearing the identical, improbable hairstyle. It bordered on parody, and the employee got sent home in disgrace.

And how much do you want to be like your boss, in any case? When Melville Bell Grosvenor served as editor and president of the National

Geographic Society in the 1960s, his staff knew him fondly as "the skipper" because of his Naval Academy background and because he was an enthusiastic sailor. Subordinates went out and bought sailboats of their own and showed off their loyalty by coming to work in Top-Siders and nautical caps. Luis Marden, a photographer and writer, also bought a boat, but commented, "It's a good thing the skipper isn't interested in fire-walking."

The urge to imitate, rather than innovate, is one of the central facts of business life, not just for individuals, but for companies. Any successful new product begets a flock of clones—the TiVo digital video recording machine, the iPod digital music player, and Viagra being among the notable recent exemplars. Some companies make a strategy of imitating their competitors. Matsushita, for instance, used to be known in Japan as "Maneshita," roughly meaning "copysonic," because its business plan was to make high-quality knockoffs of products Sony had brought to market.

Smart companies also imitate themselves, tweaking their best inventions to make them work a little better, or taking a successful idea and translating it into a completely different product line. (Behold the glory of human imagination: The spinning lollipop, or Spin Pop, begets the SpinBrush, a battery-powered toothbrush, which then begets the spinning Dawn power dish brush and the spinning Tide laundry brush. And in this case imitation clearly pays: Procter & Gamble does $200 million a year in global sales of the toothbrush alone.)

Some companies adamantly resist imitation. Using lawsuits, confidentiality agreements, and lobbying power, for instance, the drug manufacturer Wyeth has managed to protect, for decades beyond the normal patent period, its monopoly on the $800-million-a-year hormone-therapy drug Premarin, which is derived from the urine of pregnant mares.

Other companies deliberately invite imitation. IBM did it, for example, to establish its personal computer as the industry standard and is currently trying it again with its BladeCenter motherboard architecture and with the recent release of five hundred patents for use by open-source software developers.

Some ostensibly creative industries live and die by imitation. Bollywood and Hong Kong moviemakers ape Hollywood, and Hollywood in turn apes them. They all meanwhile ape themselves, with sequels and prequels ad nauseam and beyond. Successful hijackings and bank robberies beget copycat hijackings and robberies, and lucrative executive compensation packages beget copycat contracts. Two months after Jack Welch signed his perks-for-life package at General Electric, his friend and former GE colleague Larry Bossidy signed a similar deal at AlliedSignal. Six months later, Charles Knight, who voted to approve the Welch deal as a GE director, negotiated a copycat compensation package for himself at Emerson Electric, and a month after that, IBM (Knight again among the yea-sayers) gave Lou Gerstner a deal to match Welch's, with access to company aircraft, cars, offices, and apartments well into retirement. Stockbrokers, finally, do not take a random walk down Wall Street. They buy and sell what other brokers are buying and selling and thus, according to a recent analysis in the journal *Physical Review Letters*, move like a herd of wildebeest on the plains of Africa.

CAVEAT IMITATOR

As long as everyone manages to avoid plagiarism, violations of noncompete clauses, and copyright- or patent-infringement lawsuits, is there any reason to refrain from imitation? Or to feel regret about it, except perhaps for shareholders who get stuck with the tab?

Imitation is natural. It is our default mode. The trick is merely to avoid imitating stupidly. And it seems reasonable to believe that we do so by imitating success, as in the diner parking lot theory. But our natural urge to shun failure and imitate success can also be hazardous. In a paper pointedly titled "In Search of Excellence: Fads, Success Stories, and Adaptive Emulation," two Cornell University researchers blame a chronic fascination with business success stories for the tendency of corporate management theories to rise and fall in wildly faddish (not to say foolish) cycles.

"Success stories dominate business discourse to the virtual exclusion of close theoretical or comparative analyses," especially when a new theory is just starting to take off, authors David Strang and Michael W. Macy argue. "We suggest that a preoccupation with performance can paradoxically generate waves of adoption of innovations that are worthless, or nearly so, followed by waves of abandonment."

For instance, the idea of quality circles, aimed at getting assembly line workers actively involved in improving quality control, took off in the early 1980s. This was the supposed "management secret" enabling Japanese companies to outcompete U.S. rivals. Articles touted the almost miraculous success of American early adopters such as Lockheed, which boasted "costs savings of $3 million, tenfold defect reduction, a 6:1 return on investment, and 90% employee satisfaction."

During the period when quality circles were becoming a hot topic, "no articles traced quality circle failures," according to Strang and Macy. And during a flurry of adoptions that swept through the entire Fortune 500 in the early 1980s, nobody seems to have noticed that Lockheed itself had actually abandoned quality circles in 1979. By 1988, almost all such efforts in the Fortune 500 had failed or faded away.

What went wrong? Strang and Macy constructed a series of mathematical models that demonstrated, in essence, how almost any dumb idea can become the hot new management trend through imitation. Or as they put it: "Suppose that every firm uses a different innovation, all of which are entirely worthless, and that all firms have an equal shot at success in a highly competitive market . . . it is only a matter of time until, by chance, consecutive winners happen to be using the same innovation."

Observers assume that because something—in this case success—happened after the innovation, it must have happened because of the innovation. "Even if only a few daring firms copy the leaders, the odds have now been tilted slightly toward a third winner using the same innovation. Eventually there will be three consecutive wins, which may turn the heads even of the skeptics."

In the real world at this point, high-priced consultants begin to circle around the newly hatched Big Idea (formerly the worthless innovation). Strang and Macy call them the "carriers," and like the vectors of a disease they quickly spread the Big Idea around the corporate marketplace. In the case of quality circles, for instance, "consulting activities first rose dramatically, from two consulting firms with 11 full-time consultants in 1980 to 60 firms employing 469 consultants in 1983. Five years later, two-thirds of these firms had exited the market." Some consultants simply moved on to ride the next faddish wave with Big Ideas such as "job enrichment," "total quality management," or "reengineering."

Strang and Macy assume "that managers are exceptionally intelligent and under intense pressure to get it right, and that consultants charge a premium to implement hot innovations." So the faddishness of the business community is puzzling. But they conclude that "this faddish behavior may occur not in spite of these performance pressures and high consulting fees but *because* of them. Consultants advertise their winners, not their also-rans. And managers feel pressure from executive boards and stockholders to emulate peer success." The Big Idea eventually fades away as adopters find that they cannot duplicate the reported success in their own companies. But by then the original proponents have long since run for the exits and the average worker has shrugged off the whole experience and gone back to doing real work.

So if imitating success is entirely natural, and if managers face continual pressure to find new avenues of success for their own companies, how can they avoid paying a high price for the next Big Idea/worthless innovation? Strang and Macy distill the implications of their research to what they call "Montaigne's heuristic," a heuristic being a shorthand rule for choosing the best course of action in an uncertain world.

This heuristic derives from an incident described by the sixteenth-century French essayist Montaigne: An atheist visiting the Greek island of Samothrace was admiring the votive offerings left behind by those who had survived shipwrecks, and a believer challenged him to explain his atheism in the light of so many lives clearly saved by the gods. The atheist's reply was simple: Those who had drowned were more numerous; they just weren't as good about leaving behind offerings.

What Montaigne's heuristic suggests is that the next time a consultant comes calling with the latest Big Idea, "look not only at the believers who were saved, but at the even larger number who drowned. Or in more contemporary terms, look for best practice not only among the Intels and Microsofts, but also among the Wangs and Digitals."

Get the consultant to go beyond the sales pitch boilerplate: "Okay, now tell us about a couple of companies where the idea didn't work, and tell us why." Get a sense of whether past success is something anyone can emulate. Does it work only in certain conditions? Or only under the leadership of a few true believers? Imitating success may still prove to be a useful strategy, but only when it is imitation based on skeptical questioning.

Unfortunately, Strang and Macy conclude, "The trend in the contemporary business community . . . is in the opposite direction."

But maybe this doesn't matter as much as it once seemed to. Maybe the faddish brainwaves of top executives and their consultants make about as much difference, coming and going, as the wave passing through the stands of a crowded football stadium. The game goes on, regardless. Observers increasingly view the behavior of customers and employees alike less in terms of what managerial or other elites dictate and more in terms of self-organizing networks, a phenomenon most easily understood by starting with the natural world.

NO VICE PRESIDENTS

It's a winter evening by a water hole in Botswana's Okavango Delta. The hippos are lounging eyeball-deep. The sun drops earthward, painting an orange stripe across the horizon, and suddenly a river of small birds comes streaming in, tens of thousands of them packed close together, swerving and undulating like a single fluid organism. A red-necked falcon dives into the stream, to pluck out a bird for dinner, and the flock pulses out to the opposite side. These sudden evasive maneuvers occur, like all the flock's movements, in utter unison, and the birds, a species called quelea, with a reputation as avian locusts, just keep coming. The feathery heads of the reeds in the pond are lit up at first like candle flames by the last remnants of sun, but they bend toward the darkness under the weight of the birds coming in to roost.

How do they do it? How does a flock of birds, a school of fish, a herd of wildebeest, or a swarm of insects achieve that perfect choreography? Who calls the tune? Where is the dance master? Envious human observers have long assumed that such synchrony requires centralized control, some underappreciated vice president for pitch and yaw, perhaps—or, given that a school of herring can stretch seventeen miles and include millions of individuals, maybe a whole hierarchy of flight controllers, school disciplinarians, ward captains, and tail nippers. But this assumption turns out to be wrong.

The individuals in flocks and schools largely figure out what to do next on their own, without anyone telling them how or when. Scientists call it self-organizing behavior: Complex structures get built by animals blindly obeying a few elementary rules coded into their genes. Perfect harmony somehow arises out of thousands of individuals acting independently, responding to local information, often by simply imitating their neighbors. This probably should not be as surprising as it seems. What kind of hierarchy, after all, is quick enough to orchestrate instantaneous synchronized reactions to momentary threats like a dive-bombing falcon?

The problem in the past was that no one could figure out any mechanism, other than central control, to explain this synchrony. The alternative idea, that complex behaviors could happen without control, by self-organization, only began to evolve in the 1950s. Scientists started by looking at the ways simple chemical and physical reactions can create complex patterns, like the wind rippling of dunes or the array of hexagonal cells on the surface of oil in an evenly heated pan. Other researchers then extended the search for such self-organized behaviors into the animal world, beginning with ants, bees, and other social insects.

One lightbulb moment for popular understanding came in the mid-1980s, when Craig Reynolds, an expert in computer animation of complex behaviors, set out to replicate the way a flock of birds darts and weaves across the sky. Biologists had by then come to think that the individuals in flocks were responding not to some higher authority but to the movements of other individuals in their immediate vicinity. Reynolds found that he could make computerized bird-oids, or boids, demonstrate all the flight behaviors of a flock by programming them to follow three simple rules: Avoid collisions with nearby flockmates. Match their speed and heading. And stay close.

The resulting flocks were realistic enough in their banking and diving, their mercurial and yet seemingly choreographed flashes of movement, to fool even ornithologists. Among other applications, boids-style software has since enabled animated flocks of penguins and schools of fish to move, largely free of human control, in movies from *Batman Returns* to *Finding Nemo*. Over the past few years, self-organizing behaviors have also turned up everywhere in the natural world.

SELF-ORGANIZED SKYSCRAPERS

Termite mounds are among the most astonishing products of self-organization. They stand, some of them fifteen feet tall, like weird druidic monuments in the African landscape. Each mound rises from a broad pyramidal base, then narrows into a finger of clay, slightly bent, as if beckoning to the sky. If termites were the size of humans, their largest mounds would be triple the height of our tallest skyscrapers and almost as complex.

Deep within the mound, the central chamber is like the assembly plant of some demented genius in a James Bond movie. Pale quarter-inch-long workers bustle everywhere, and soldiers with big copper-colored heads and amber abdomens rush out to defend the nest, snapping their curved mandibles. The chamber contains a half-dozen clay shelves, where orange, fibrous structures like honeycombs stand on stubby little feet. It's a garden without sunlight. The termites grow a fungus here and use it to predigest the tough, dried grass they bring in for food.

These incredibly complex structures, each housing a small city of more than a million individuals, enable the termites to dominate the African wilderness around them. The termites eat more grass than all the wildebeest, Cape buffalo, and other savanna mammals put together. But the real wonder is that there is no demented genius running the show. The termites do it all without any blueprints or supervisors.

How does a termite mound get started? Put termites in a dish lined with an even layer of soil, and they go through what a French researcher termed *"la phase d'incoordination,"* with workers randomly piling up grains of dirt and taking them down again, and nobody quite sure what they're doing. But eventually enough pellets accumulate in one place and everybody piles on in a coordinated effort.

The termites still don't know what they are doing or where they are going. But the workers cement grains of sand into place with their saliva, and the saliva contains an attractive pheromone, which causes other workers to come and add to the pile. This positive feedback loop builds into a construction frenzy. Columns of dirt rise, expand into walls, and get roofed over, and the mound gradually begins to form. The actual shape of different structures arises spontaneously, not from any blueprint, but from the influence of physical and chemical factors, as well as the size of the termites' own bodies.

If an intruder breaks through the wall of the mound, the change of atmosphere causes workers to rush forward and pile up pellets of dirt on the damaged wall. As the tunnels get capped off, the concentration of the workers' own attractive pheromone builds up and the termites pack the area with dirt. A week or two later, the wall of the mound is as solid as if no disturbance had ever occurred, all without any boss ever having said, "Fix this problem."

DEMOCRACY IN ACTION

Even in species with clear hierarchies and relatively high intelligence, animals seem to have a surprising knack for self-organization. Some researchers call it democracy. For example, red deer generally move not when the alpha stag says so, but when roughly 60 percent of the adults "vote" by standing up and looking restless. Whooper swans signal the urge to fly with movements of their heads, and the group actually takes off when these signals rise to a certain threshold of intensity. Even gorillas generally decide to move based on the calling of a majority of the adults in a group. (So much for that legendary autocrat, the eight-hundred-pound gorilla.) These species all have clear hierarchies. But research by Larissa Conradt and Tim Roper at the University of Sussex suggests that it's often just too costly for dominant individuals to impose their will on the group. High-ranking animals may sometimes manage to steer group behaviors in a particular direction, or they may exploit the behaviors of a group, but they seldom dictate what those behaviors will be.

And in humans? We are not flocks of penguins or schools of fish. We can think about what we are doing and continually invent better ways of doing it. But we also self-organize in certain circumstances—for instance, with some of the imitative behaviors, such as phantom traffic jams, described earlier in this chapter. It's also a form of self-organization, Tim Roper suggests, when people at a meeting begin to rustle their papers or place their hands flat on the table. The message is as plain as the voting of a herd of red deer: *This meeting has gone on too long.* Managers ignore such unspoken sentiments at their own considerable peril.

The delicate balance between hierarchy and self-organization typically shifts as an organization grows in size. A small paper wasp colony with a population of a dozen or so individuals is likely to be tightly controlled and hierarchical. The queen knows all of her subordinates and can examine the entire nest to figure out what they need to be doing next. But larger colonies tend to become less hierarchical and more self-organized. The queen can't be everywhere. The business of policing the nest shifts from the queen to the workers themselves.

The same thing happens naturally in the human workplace. In the early years of Southwest Airlines, founder Herbert D. Kelleher was the soul of the company. He "loved to joke with employees and dole out hugs and kisses as greetings," according to the *Wall Street Journal.* "Original employees—many of whom say they considered him more of a father than a boss—wanted to please and emulate him." But by the time he retired in 2001, the company had thirty-five thousand employees. Company workers now policed one another, challenging questionable sick calls, overuse of office supplies, and requests for overtime pay for "stupid things."

When people talk about self-organization in the human workplace, they generally mean that individuals choose how they participate and collaborate together. The hierarchy no longer attempts to control individual behavior. Workplaces built on this model look less like a symphony orchestra, in which each player performs an assigned part by rote, says Cornell sociologist Michael Macy, and more like an improvisational jazz group, in which the players find their own way through the music and yet achieve something together that is more than random noise. "You really have to pay attention to what the other artists are doing—you have to fit with what they're doing and yet not be doing exactly what they're doing." The venerable forces of imitation, synchrony, and emotional contagion are once again in control.

The hierarchy serves mainly to establish loose rules of play. Scientists who study self-organization talk about "tuning parameters"—that is, making subtle changes that produce a "bifurcation," a shift to a strikingly different self-organized pattern or behavior. In the natural world, the parameters tend to get tuned by environmental factors. When the temperature drops below a certain point, for instance, a bifurcation occurs, and water suddenly becomes ice. But in human groups, the environment often gets created or modified by managers. For instance, traffic managers are tuning the parameters when they raise the toll on a busy highway during rush hour, producing a sudden shift in traffic patterns. Architects are tuning the parameters when they design a workplace that encourages a high level of social contact, or of isolation. Our behaviors self-organize, often unconsciously, around these new parameters.

Some of the most successful new businesses of the past decade have exploited the Internet to create what appear to be self-organizing networks. Amazon, Apple iTunes, and Netflix all attempt to create communities of opinion, with ordinary consumers reviewing books, songs, and movies and thus helping to steer purchasing by other consumers.

Self-organizing networks have also demonstrated their ability to create and improve products even when no single company is in control. The most celebrated example is the Linux computer operating system, which Linus Torvalds invented and then put up on the Internet for anyone in the open-source community to modify and improve. In a 1991 e-mail Torvalds downplayed Linux as "just a hobby, [it] won't be big and professional."

Linux now outcompetes Microsoft in the market for Internet server operating systems, and it drives a vast array of so-called embedded devices in products such as automatic teller machines, TiVo recorders, Linksys wireless routers, and personal digital assistants. The Linux self-organizing network consists of volunteers around the world, motivated largely by the old tribal social forces of prestige, trust, and reciprocity. The *Harvard Business Review* recently celebrated it as "the first and only market force capable of challenging Microsoft's hierarchical approach to software development."

Self-organizing networks and communities also flourish independently of the Internet. It is basic human behavior, as well as basic animal behavior. Such informal, highly collaborative efforts typically characterize any new start-up, according to Kathleen Carley, who studies self-organization as a professor at Carnegie-Mellon University. They're also the rule in small contract-dependent companies, which can shift direction almost as suddenly as a flock of birds, particularly in sectors such as defense and health care. One or two employees kick around an idea together, decide that they can get these three other people working with them on something productive, says Carley, and then they go out to seek a grant. Suddenly the company is hiring people to work on the new grant and may even find itself shifting in a totally unanticipated direction.

By way of example, she cites BBN, the Cambridge, Massachusetts, research and development firm, which started out doing acoustical consulting. But it went on to create ARPAnet, forerunner of the Internet, and to send the first person-to-person network e-mail. More recently, it developed a system to create real-time transcripts and indexing of any audio source, in Spanish, English, Chinese, or Arabic, using standard desktop hardware.

Even within some large corporations, managers have begun to think of their work less in terms of command and control and more in terms of facilitating the informal, self-organizing behaviors of employees. David Ticoll, a Toronto business strategist, cites Harrah's, the casino company, as an example. It now distributes minute-by-minute information on the status, whereabouts, and gambling behavior of VIP guests, so frontline workers can bestow perks and otherwise coddle good customers at their own discretion, albeit "guided by well-defined rules and performance objectives."

Likewise, when British Petroleum set out to reduce greenhouse gas emissions in the late 1990s, it avoided conventional top-down pollution management and instead created an internal marketplace where business units could buy and sell pollution credits at their own discretion. Using self-organization enabled BP to surpass its greenhouse gas emissions targets at reduced cost. The company is currently shifting into external pollution credit trading markets, where more players make for more opportunities. But the principle of self-organization remains the same.

To Ticoll, these cases suggest the emergence of a new business model, "the tightly controlled, hierarchical production or distribution system with features of chaotic self-organization." He argues, for instance, that Amazon, Apple iTunes, and Netflix all "shrewdly facilitate self-organizing exchange" but that "all three have brilliantly designed the customer experience from end to end, even more than typical hierarchical retail stores."

Carley points out that self-organizing techniques do not suit some companies. They scare managers because they can involve a loosening of control, particularly so in the technology sector, where lines of power and lines of information are often tightly connected. An open, collaborative environment, where everybody has access to everything, is also an environment where any employee can walk out the door with vital intellectual property. "It's not all good," she adds.

But she also says that self-organizing behavior has "captured the public relations imagination." Technology has made it possible for proponents to "model it, to simulate it" with the sort of dazzling graphics that wake up otherwise inattentive executives. Business consultants are already beginning to tout the idea.

In other words, self-organizing networks might just become the next Big Idea. Imitators ought, as always, to beware.

> When he was a psychologist at Yale in the first half of the twentieth century, Robert M. Yerkes equipped his chimpanzee holding areas with a push-button water fountain, like the one in any workplace. He worried at first that the staff would have to teach each animal individually how to use it. But in fact, even newcomers picked up the technology merely by watching and imitating other chimps. Yerkes described the events at the water fountain as an instance of the cultural life of chimpanzees.

But it seems somehow grander than that, doesn't it? The moment when that hairy finger first touched the button on the office water fountain demands Wagnerian music, *fortissimo,* as if to announce the dawn of the corporate primate.

Soon after, a group of chimpanzees sat down together and invented the coffee break.

13

BUNNIES FOR LUNCH
On Being a Corporate Predator

I have a bunny. And it is one of the original ones. But I don't like it. I can't help it—I always gravitate to the predators. It's just my nature.

—PUBLISHING EXECUTIVE PETER OLSON,
ON HIS COLLECTION OF STEIFF ANIMALS

Oh, to hell with the neuroeconomics and the evolutionary psychology.

The truth is that you just want to be a goddamn predator. You want senior executives to tremble when you pass. In your heart, your Steiff animal is the man-, woman-, and child-eating leopard of Rudraprayag. Your fondest wish is to be an equal opportunity monster. *Grrrrr.*

Friend, I can help you with that. But do you really need to become a cliché in the process? Corporations spend millions developing eloquent, iconic logos and otherwise cultivating their image, and then their executives go around prattling about being lions, foxes, tigers, and sharks. It's so pedestrian, like having lunch in the Grill Room at the Four Seasons and asking for ketchup with your bison carpaccio.

Would-be predators ought to show a little flair, shouldn't they—a certain ruthless urge to take their strips of raw, bleeding meat *au naturel,* in the manner of the war correspondent who after a tour of battlefield carnage adjourned with a queasy newcomer to his favorite Saigon restaurant and blithely ordered *"un steak tartare et un vin rouge"*?

Well, isn't that the idea? Don't really memorable predators

display an unwholesome quality of delectation, like Hannibal Lector with his fava beans and Chianti?

This is probably going to come as a dreadful disappointment to the excitable boys in the executive boardroom, but it isn't like that in the animal world. Even for really fierce predators, getting a living is a highly methodical business. Animals can, of course, teach useful lessons about carving out a few beating hearts, a few glistening livers. But the lessons are less about being bloody and bad and more about being shrewdly opportunistic. Let's begin by dispensing with a few myths about popular models from the animal world.

- *You don't want to be an eight-hundred-pound gorilla.* In fact, no such animal has ever existed. The average big daddy silverback tops out at about half that legendary weight. And gorillas are not predators, in any case, but vegans, with an almost unlimited appetite for fruit and bamboo shoots.

I once worked on a Discovery Channel documentary about lowland gorillas where the dramatic episodes on an average day consisted of the alpha male farting, picking his nose, and yawning. Then he did the same things, the other way around. Over and over. A lot like some offices. But this is probably not the image you want to present to the public. It's surely not what a top executive at the BBC had in mind, for instance, when he described his organization and Rupert Murdoch's BSkyB as the two eight hundred-pound gorillas of British broadcasting.

- *Hold that lion.* Nor do you want to model yourself indiscriminately on lions, though you may pride yourself on your roar or your bite. Once, traveling in Botswana, I saw a male lion rouse himself to court a female, with lots of growling and nipping. Finally, like a tired wife, she assumed the sphinx position for sex, and he mounted her. One of my companions, a *National Geographic* photographer, began whirring and clicking (with his camera, I mean—though on these long road trips one can never be too certain). The big moment of leonine love lasted all of ten seconds. "Definitely a motor-drive picture," the photographer muttered.

Think about this and nod appreciatively, even admiringly, next time the preening CEOs at an awards banquet liken one another to lions. And remind yourself that in the modern workplace, much as during our evolution in the Serengeti, it really does pay to know your natural history.

▪ *Piranhas are pussycats.* The piranha is another common model for cutthroat corporate behavior, as when the *Wall Street Journal* extolled Canadian takeover artist Gerald W. Schwartz for "stalking deals like a piranha after a goldfish." But piranha predation is seldom the glorious, gruesome bloodbath of popular lore, and no one has ever documented a case where piranhas have actually killed a human being.

In the course of various assignments, I have climbed into a tank of hungry piranhas at feeding time, I have swum with piranhas in the Rio Negro, and I have caught and released piranhas while standing hip-deep in the Amazon, all without so much as a nibble.

The most dangerous animal I encountered along the way was a budget-driven assistant producer for National Geographic Television, who had bought me a bright red bathing suit to film the tank sequence and practically tore it off me afterward so she could take it back for a refund. It was apparently defective on the grounds that I was not dead.

Piranha swarms are in fact a case of popular delusions and the madness of crowds. Swarming generally occurs in just two circumstances: at bird rookeries and fishing docks, where the piranhas gang up to fight it out with one another because a steady supply of food is hitting the water, and on flooded plains where the falling waters have stranded them in ponds that are rapidly drying up. In the first case, they risk accidentally becoming other piranhas' dinner. In the second, everybody is about to die anyway. The moral is to be careful when you follow the crowd and always keep one eye on the exit.

What smart piranhas mostly do is snatch their food on the sly. They lurk in the shadows, nip out to bite their victim on the tail, then dart away again. One bloody-minded researcher has argued that attacking the tail fins is the equivalent of wolves severing the hamstring of a deer, to cripple the prey for an easier kill. But the truth is that piranhas seldom actually proceed to the kill.

A piece of tail fin, or a section of scales raked off like roof shingles, is all they are really after. Scales and fins can be up to 85 percent protein, and—this is the glory part—*they grow back.* So the piranhas get to lurk around the neighborhood and do it again a few weeks later.

Some piranhas even disguise themselves and tag along in a school of another species. Now and then they nip off the tail of an unwary neighbor, then swim along looking innocent again.

Would this milder sort of piranha strategy also work in the human marketplace? Of course. It's commonplace, as when PayPal skims off a few pennies from every eBay transaction, for instance, or when investors profit on arbitrage differences in the prices of stocks or bonds in different markets. It's not about being a monster. It's about being shrewd, and maybe a little sneaky.

It can, of course, also sometimes be illegal. Managers at some Taco Bell restaurants and Kinko's copy centers have been found guilty of cutting costs by erasing hours from employees' work-time records. Workers have also charged managers at Family Dollar, Pep Boys, Wal-Mart, and Toys "R" Us with the practice, commonly called "shaving time."

■ *Be a sit-and-wait predator.* Mark Twain once wrote, "The spider looks for a merchant who doesn't advertise so he can spin a web across his door and lead a life of undisturbed peace." But Twain was mistaken about a spider's appetite for peace. A spider needs to eat about 15 percent of its own body weight daily. Imagine the human equivalent as your new job description: *The ideal candidate will catch, kill, skin, and prepare four or five rabbits a day, every day, without a gun.* You might just be able to do it by chasing after your prey, if you were sufficiently fast and more than a little stupid.

Most spiders prefer to build a web and let dinner come to them. Not only is the sit-and-wait strategy smart, but spiders make it smarter by learning to build their webs asymmetrically, with more catching surface on the lower half, where it's easier to get to their prey. (Why run up and down stairs needlessly?) Happily for us, this strategy works. A British researcher has calculated that the insects consumed by spiders each year nationwide would easily outweigh the nation's human inhabitants.

Can the sit-and-wait strategy also succeed in the human workplace? The Google Internet search engine operates like a spiderweb. The prey—that's you—types in a search term indicating an interest in, say, central heating systems or naked Hollywood starlets. In response, "sponsored links" turn up on the side of the computer screen from manufacturers and retailers hoping to lure you into their web. Unlike some other search engines, Google doesn't let the sponsored links taint the editorial search results on the left side of the screen. The distinction increases user confidence about entering either side of the web.

By contrast, trying to drum up business by sending out spam e-mail is a case of chasing rabbits without a gun. You irritate the rabbits and, more often than not, end up hungry and with your face in the mud. (This may be why spammers often appear to be desperate rednecks trying to claw their way up into a better class of trailer park.)

Not to pick on Mark Twain, but another instance of the sit-and-wait strategy suggests he may also have been wrong about the universal value of advertising: Some retailers let big companies do the heavy advertising, then simply place their lower-cost store brands next to the name brands on the supermarket shelf, where they do not have to wait long for discerning shoppers to come to them.

Even better, some mom-and-pop businesses are learning to exploit the national chains instead of simply lying back and getting run over by them. Facing competition from Lenscrafters, Pearle Vision, and Target, for instance, a Lafayette, California, shop called Art and Science of Eyewear has managed to survive by, among other tactics, moving to a new location between two busy chains, a Starbucks and a Cold Stone Creamery. Says owner Anna Fuentes: "These guys are smart, they do their market research. We piggybacked on that."

FORAGING THEORY

The more you watch animals, the more you realize that survival isn't about being a mover and a shaker. Animals generally don't spend ten or twelve hours a day working for a living. They're too smart for that. Lions, in fact, loaf for up to twenty hours a day and sometimes let perfectly good prey graze a few dozen yards away without even lifting their heads to say boo. What looks like idleness is often a case of balancing risks and rewards and making prudent choices.

Over the past thirty years, biologists have increasingly relied on economic theory to understand the choices animals make as they forage for their daily meat. It is a little disconcerting at first to see animal behavior described in terms of per capita growth rate, statistical decision theory, or the asset protection principle, or to read a sentence like this: "When foraging, animals should and do demand hazardous duty pay."

But biologists and economists alike have discovered that animals aren't dumb when it comes to decision making. They aren't just empty maws cramming down whatever food happens to turn up in their path. They may actually be better than we are at choosing the best course of action for a given circumstance, because animals, unlike most corporate executives, are accustomed to operating without a safety net.

Even a humble shrew must make highly sophisticated judgments about the risks it is taking and the likely rewards, and the shrew has to get these things in balance more often than the average MBA for an elementary reason: When an MBA screws up, he's only losing other people's money. When a shrew gets the risk-reward equation wrong, he's probably dead. This is a highly effective teaching mechanism, and one that should perhaps be used more often with MBAs. It makes animals acutely sensitive to the nuances of the marketplace.

In 1986, biologists David W. Stephens and John R. Krebs brought together the various economic approaches to animal behavior in an influential book called *Foraging Theory*. It was not written for the math-phobic reader, present company included. The simplest equation in the book expresses the hunting behavior of a fox, say, as

$$R = E_f / (T_s + T_h)$$

where R is the rate of intake, E_f is net energy gained, T_s is the time spent searching and T_h is the time spent handling the prey.

This is a complete recipe for terror in a field mouse. It also suggests that survival in the natural world, like survival in the business world, depends on keeping a close eye on costs.

The book has become a manifesto for the study of how animals forage for food, and this in turn has helped illuminate some of the unexpected biological underpinnings that unconsciously shape human getting-and-spending.

Krebs and two of his zoological colleagues at Oxford University, Alex Kacelnik and Edward Mitchell, later launched a consulting business, Oxford Risk Research & Analysis (ORRA), to apply ideas from foraging theory and experimental economics to the commercial world. In one case, the study of how starlings forage for worms led directly to a revised analysis of how a British energy conglomerate searches for new places to drill oil wells. What the starlings taught the geologists and engineers helped in formulating a new approach to managing the conglomerate's portfolio of oil wells, with a projected profit difference, by ORRA's estimate, of U.S. $519 million over five years.

This is a big leap for an animal group that has previously taught us only a somewhat dubious lesson about early birds catching the worm. Kacelnik explains it this way: "A starling can choose a 'fixed' patch in which it can be fairly certain of getting a reasonable number of worms, or a 'risky' patch in which it could get a lot of worms but could also find nothing. Which should it go for? An oil industry executive, when he heard how we go about modeling such decisions, said, 'Hey, this sounds just like the sorts of problems we face in looking for oil.' "

The Oxford biologists agreed to sit down with the oil company and explore how biological ideas might help the company think differently about risk. For example, both the bird and the oil company want to make the best possible choices in the face of considerable uncertainty. When Kacelnik and Krebs were first studying titmice and later starlings, ideas about optimal foraging behavior suggested that the birds would consistently choose the safe, steady option. Common sense likewise suggested that the oil company would concentrate its exploration in areas known to produce the highest yields.

But in fact, the birds tended to combine exploitation of the fixed patch with exploratory ventures into the risky patch, apparently as a way of monitoring the shifting probabilities of success. Such sampling forays make sense because starlings and other animals live in a world where the odds are constantly changing "There is a certain value to be gained in increasing knowledge," says Kacelnik, "even if you are not getting the immediate maximization of payoff."

The oil company had apparently forgotten this lesson. Exploration teams tended to be too cautious about going into unfamiliar regions. On the other hand, they tended to overestimate the probability of success in regions where they had already developed a level of comfort. They drilled wells at a cost of $20 million apiece in familiar areas when a more realistic estimate of the available reserves might have told them the wells would never be commercially viable.

Researchers at Xerox have also begun to explore commercial applications of foraging theory, as a tool for developing better ways of finding information on the Internet. They argue that the failure of Web sites to accommodate our evolved foraging behavior is one reason roughly 65 percent of Internet shopping trips end in failure. But before we get to that, let's consider some of the simpler lessons foraging theory can teach:

- *Pick the right prey.* In one of the classic experiments in foraging theory, John Krebs studied a European species called the great titmouse, placing individual birds in a cage with a small opening in the bottom. A conveyor belt carried big mealworms and small ones past this window. The small ones were half the size of the big ones and also had a strip of tape attached, which the bird had to remove before eating. Some birds needed as little as five seconds to handle a small mealworm, while others took almost twice as long. The more adept birds could afford to eat both large and small prey, regardless of the speed of the conveyor belt. But at a certain point, when the big worms started to turn up roughly every 6.5 seconds, the cost of handling made small worms unprofitable for slower birds, and they got out of that line of business.

Being sensitive to opportunity cost is apparently a basic animal behavior: You shouldn't attack a prey or pursue an opportunity if it's going to distract you from pursuing a better opportunity. That doesn't necessarily mean bypassing small-potato opportunities. Even the slow birds go back every now and then to sample a small worm, just to make sure they aren't missing something. But it means constantly assessing your own abilities and the nature of the opportunities rolling past your window.

For a huge company such as General Electric under Jack Welch, picking the right prey meant focusing on large, high-growth markets where GE could be one of the top two players. On the other hand, smaller companies with lower costs, or more modest expectations, could live well on GE's castoffs.

For an animal, picking the right prey is a matter of fundamental questions such as how easy it is to find, kill, and consume, how well it fills the belly, and how hungry the predator is to start with. People in the developed world tend to get distracted by less visceral standards of success, so we often end up focusing on the wrong prey.

When "Chainsaw" Al Dunlap arrived at Scott Paper in the early 1990s, for instance, he may not have done much else good, but he identified a clear case of an entire company in pursuit of the wrong quarry. Managers thought they were in the business of selling paper, a commodity subject to severe cyclical price fluctuations: "When I talked about marketing, they talked about tons. . . . Whenever I brought up profits, somebody inevitably would say, 'Well, we sold this many tons.' " Scott was so obsessed with tonnage that it routinely sold its paper for generic store brands, when it could have been earning double the profit pushing its own Scott products.

In a sense, from their individual perspective managers were pursuing the right prey. Tonnage was safe; it made quarterly targets achievable. It kept the machines running, the workers employed, and everybody happy—except the shareholders.

In the same vein, a recent study at a high-tech services company found that its salespeople were often chasing small accounts because they were easier to catch and handle. But careful tracking of sales via laptop computer makes it possible to calculate the actual costs and rewards of this approach. So if the sales manager had been paying attention to opportunity cost, as a lion would do, he would have discovered that the large accounts paid off at a rate of $2,000 in revenue per sales hour, double the yield of small accounts.

In this case, it may not pay for the company to lift its head and say boo every time a small opportunity passes by.

- *Pick the right patch.* Grizzly bears in Yellowstone National Park for years fed only on garbage dumps and tourist handouts. It was a soft life. People thought the bears had gotten too fat and lazy to do anything else. So when the Park Service decided to cut off human sources of food in 1970, critics argued that the bears would starve. And in fact, the Park Service wound up killing many bears that continued to troll for Twinkies.

But the survivors fell back on their dimly recalled repertoire of tricks for exploiting the natural marketplace. Instead of doing the same thing over and over, they moved from one patch to the next, pursuing new opportunities according to changing seasons and circumstances.

They seemed to calculate where they could make the greatest profit with the least expenditure at any given moment. In wet spring weather, for instance, earthworms congregate under tufted hairgrass and the bears go around flipping over the tufts and lapping up the worms. In trout spawning season, the bears spend their time around rivers and creeks, scooping up a daily bag of a hundred or so cutthroat trout. In elk calving season, they bird-dog through the sagebrush looking for spots where Mama Elk has bedded down Bambi for the day. (Why not dine on Big Mama instead? This is foraging theory in action, and another case of picking the right prey: Adults are relatively fast and get protection from the herd, so a grizzly may chase after them all day in a vain search for one big meal. But fawns can't keep up with the adults, so their survival strategy is to lie absolutely still and hope a grizzly doesn't find them before Mama comes back late in the day. The bears size up the odds and set out to find the babies. And in this case, the strategy of pursuing lots of small meals with a high probability of success seems to pay off, at least for a week or two each June. In a field at Yellowstone, I once saw a bear kill five fawns in little more than an hour.)

When the baby elk patch plays out, bears shift to humbler fare. In August, for instance, they head up into the mountaintops, where they have somehow discovered pockets of estivating army cutworm moths to scoop up for dinner.

What does this have to do with workplace behavior? Humans, like grizzly bears, must sometimes also manage the shift from fat and lazy to lean and flexible. Before the pickings get slim in the patch we've been working, we need to make the difficult leap to a profitable new patch.

For example, in the early 1990s as the personal computer era was taking off, IBM was still stuck in the lucrative but rapidly aging System/360 mainframe business model. For Lou Gerstner, the new CEO, the hardest challenge was taking a workforce accustomed to easy success, and largely immune from normal competition, and trying to make it "live, compete, and win in the real world. It was like taking a lion raised for all its life in captivity and suddenly teaching it to survive in the jungle."

Gerstner actually held meetings with staff at which he showed them photos of the enemy (those notorious predators Bill Gates of Microsoft,

Scott McNealy of Sun Microsystems, and Larry Ellison of Oracle) and read quotes in which they gloated over IBM's decline: "IBM? We don't even think about those guys anymore. They're not dead, but they're irrelevant." It was like holding a seminar for tabby cats and showing them photos of man-eaters. IBM still makes mainframe computers, but it is now the world's largest consulting company, with the ability to shift its focus in tandem with the needs of information-system customers.

Other companies get themselves stuck on a more or less lucrative garbage dump. Polaroid was so focused on its instant-imaging photo technology that its managers in the mid-1990s completely missed the coming digital photography revolution. Thus Polaroid went from the Dow Nifty 50 to Chapter 11 in a few decades (though it still managed to award its dim-witted senior executives and directors $6.3 million in bonuses and other fees as it was stumbling into bankruptcy court).

Chrysler, GM, and Ford currently get almost all their profits from SUVs and big pickup trucks, another tempting garbage dump. Demand was so strong in 2003 that GM built an extra two hundred thousand trucks on overtime wages, at the same time that it was laying off workers at other plants where it had built two hundred thousand unwanted conventional cars. Moreover, by contract, the furloughed workers were still getting 90 percent of their wages.

A year later, when gas prices surged, demand for SUVs suddenly flopped. But Detroit's Big Three automakers kept cranking out SUVs even when they could unload them only by offering rebates of $5,000 apiece. Expensive equipment, union contracts, and single-minded management fixation on a good thing made Detroit inflexible in the face of rapidly fluctuating demand. Japanese automakers didn't get stuck in the same patch in part because they hadn't been as successful with SUVs in the first place.

But beyond that, Toyota builds 80 percent of its vehicles on flexible assembly lines, versus 30 percent at Ford and Chrysler, and these lines can shift from one model to another in as little as a weekend. Nissan has set itself the goal of being able "to build anything, anytime, in any volume, anywhere, by anyone," meaning that it will be able to shift production almost instantly to match demand around the globe.

Learning to be flexible, to identify the right prey, or to be sensitive to the shifting value of a particular patch is seldom easy. But it is becoming a question of survival in the developed world, as it always has been in less developed nations. With good jobs getting outsourced to the Third World, you could easily imagine yourself in the position of a Yellowstone grizzly getting chucked off the garbage dump: Living fat and easy is no longer an option. The safety net is gone.

And, oh yeah, the park ranger is coming after you with a gun. So. Are you smarter than the average bear?

- *Know when to move on.* In fact, animals generally don't need park rangers to tell them it's time to give up on a good thing and seek new opportunities. Studies have demonstrated that they routinely assess a patch in terms of whether the amount of food they're getting for their effort is going up or leveling off.

That is, animals are sensitive to what economists call marginal value. If it's a long way to the next patch (like the Yellowstone grizzly's trip up to the mountaintops for the cutworm moth festival), then an animal will probably stick with what it's got a little longer, despite diminishing returns. If the next patch is nearby, or if a rival is making the current patch dangerous, then the animal will arrive somewhat sooner at what biologists call "giving up time" or "giving up density."

In his travels, Joel Brown, a biologist at the University of Illinois at Chicago, sometimes uses giving up density to gauge local poverty. People in prosperous areas often give up on a chicken bone when there's still meat on it, because they know there will be plenty more chicken legs down the road. In more parlous circumstances, people bite the cartilage off the end and strip the bone bare. And in dire poverty, they break open the bones and suck out the marrow.

Giving up density also unconsciously shapes the way we shop, which is at heart a foraging activity. We head for the mall, or a strip where a lot of car dealers are clustered together, so we can reduce the cost of traveling from one patch to the next. We go from dealer to dealer comparing prices until we reach a point of marginal value where the information we're likely to get no longer justifies meeting another car salesperson. And then we bite.

The computer has dramatically shortened this process, but our basic foraging behaviors remain the same. Researchers at Xerox's Palo

Alto Research Center (PARC) in California believe that people search for information on the Internet the way animals hunt for food: We try to find a productive patch, work there to a point of diminishing returns, then jump to another promising patch, or give up and start over again with a new search engine.

One business school professor, for instance, now does all his car shopping in a single productive patch, www.autobytel.com, which allows him to compare the available models in his area and the prices different dealers are offering. Because he is acutely aware of the value of his time and also the importance of territoriality, he then calls the dealer of his choice, gives his charge card number and the price he will pay, and tells the salesperson to deliver the car if he wants the sale. If the salesperson shows up with extra charges tacked onto the contract, the business professor shrugs and says "Sorry, that wasn't the deal." He can afford to walk away because his total investment—time spent searching plus time spent handling—now adds up to all of fifteen minutes.

Misunderstanding our evolved foraging behaviors can be enormously costly in the Internet era, according to PARC researchers Stuart Card, Ed Chi, and Peter Pirolli. American workers with Internet access spend almost half their workday online—eighteen hours a week in 2003. With roughly fifty-five million Web sites to choose from, and more than 500 billion documents, it's easy to get lost or distracted, or to be misled by unreliable information. From the point of view of the seller or Web site provider, it's also much easier to lose potential customers, because they incur so little cost in getting to a Web site and can thus afford to give up and go elsewhere at the slightest frustration.

The PARC researchers describe the search process largely in terms of "information scent." Say a benefits analyst asks her favorite search engine to find "alternative cancer cures." The search produces a list of relevant Web sites, and the keywords that turn up in each site description serve as scent cues telling her whether it's worth traveling to that site. In this case, many of the Web sites give off the wrong scent; they seem unreliable or even fraudulent. So, as a quick way to get herself into a more productive patch, the benefits analyst decides to skip commercial sites ending in *.com* (by adding "-.com" to her search term) and focus only on university sites ending in *.edu.* Or she may choose to ignore sites that mention "apricot seeds" or "laetrile" and follow a scent trail that includes the keywords "NIH experiment."

The importance of "information scent" is now widely accepted among Web site designers. It has given the industry "a whole new way of thinking about things," says Jared Spool, of User Interface Engineering, a Massachusetts research company focusing on Web site design. "Up until we started talking about information scent, nobody had a way of describing why some sites feel better than others. They just did. Now we can turn to information scent and say, 'Well, *look*.' "

But Spool adds that foraging theory hasn't yet provided the tools to enable designers to deliver Web sites with good information scent. PARC researchers have experimented with two software programs in a search for ways to deliver those tools. ScentTrails keeps track of steps a user takes in the search process and uses this information to scan ahead for traces of that scent in potentially relevant Web sites. Then it highlights different links with greater or lesser intensity according to their relevance.

PARC's Ed Chi describes the idea this way: "Instead of chopping down all the trees in the forest until there's one tree left with all the fruit on it, which is what the filtering mechanism on a search engine does, why not just tie ribbons around the trees with fruit on them?" Enhancing the scent trail in this fashion speeds up the search process, Chi says, by more than 50 percent.

Bloodhound, another experimental software program developed at PARC, uses foraging theory to test Web sites for navigation problems in getting visitors where they want to go. Large companies typically hire consultants to conduct such tests, which can cost up to $30,000. Bloodhound does the same thing automatically, and it can test a site over and over to see how small changes affect a site's usability. Bloodhound starts with a search term, the equivalent of a T-shirt or other personal item wafted under a hound's nose. Then it sniffs its way through the Web site to get to its quarry.

Intelligently designed Web sites (for instance, www.landsend.com) cluster their offerings in a logical way and use good, informative keywords to provide visitors with clear scent trails leading to different products or services. Badly designed Web sites (for instance, www.macys.com) create olfactory confusion; they lack clear scent trails. A retailer, for instance, may put too many products on its home page, causing the overwhelmed visitor to jump to another retailer's Web site. Or the scent trail may peter out, causing Bloodhound to lie down and moan in dismay and the customer to go elsewhere.

But this is a sorry pass, to have all our rich and bloody history as predators come down to the business of clicks on a computer mouse and the hunt for pixels of information scrolling down the screen. It isn't what businesspeople generally have in mind when they talk about being predators, is it?

What they still have in mind—and no amount of foraging theory or real animal behavior is going to change this—is scaring the piss out of people. The impulse is strong. So let's at least play devil's advocate for a while, and ask whether in some cases using fear might also be strategically useful.

In a landscape populated by big, scary predators, you need to be hypervigilant to stay alive. One neotropical tree frog lays its eggs in the relative safety of the foliage alongside ponds. The longer the young can put off the moment when they must hatch and drop into the water, the better their chances for survival. Having a little more time to develop makes them more adept at dodging the jaws of aquatic predators. But lingering too long in the hatchery can also be dangerous, as climbing vine snakes sometimes come to play with baby.

The tree frog's answer is a form of prenatal vigilance. Even in the egg, embryos can recognize the vibrations produced by a feeding snake. So as the first few eggs are being lapped up, their siblings hatch prematurely and drop into the water to make their escape.

14

A LANDSCAPE OF FEAR

Why Do Jerks Seem to Prosper?

If you hit a pony over the nose at the outset of your acquaintance, he may not love you, but he will take a deep interest in your movement ever afterwards.

—RUDYARD KIPLING

One day a few years ago, the chief executive at a prominent retailer was off inspecting stores in North Carolina. The CEO, then in his mid-fifties, was recovering from a facelift but, characteristically, disregarded his doctor's orders not to travel. His face was smeared with Vaseline, there was inflammation around the corners of his eyes, his freshly stretched skin was a sickly yellow, and he was pissed. Pissed at the floor set, the way the merchandise was arranged, the color of the item at the front of a stack of shirts. "He was yelling at people," says a former employee, "screaming, 'It's the wrong color! It's making me sick. It's putrid.' "

The young workers were confused and appalled. They stared at their leader's face. Yes, they thought, the color. And scrambled to change it.

"It didn't matter what it was," the former employee recalls. "He could walk in next day and put that same color up front."

The CEO was the sort of corporate leader often admiringly described in the business press as hard-charging. Since taking control, he had transformed the company from a fusty relic into a hot retailer of clothing for the college crowd. He had almost quadrupled the number of stores while consistently producing earnings growth. A management textbook featured his methods.

No one would suggest that the CEO had achieved these results by being nice. Words such as *driven* and *determined* came more readily to mind. He had definite ideas about how certain things ought to be done. For instance, he followed a ritual every morning, always parking his car perpendicular to company headquarters, then entering the building through one revolving door, discarding a piece of paper inside, exiting by another revolving door, and reentering by the first door. He wore lucky shoes around the office, a pair of suede mocs so old and decrepit that his feet stuck out. He took notes with a lucky pen, which leaked ink on his hands. He required employees to wear the company's own products, but nothing black. Employees who showed up in black risked getting sent home after having the company's standard motivational speech shouted at them: "What are you, fucking stupid?"

Sales meetings started at 8 a.m. Mondays and sometimes lasted till midnight, in part because employees who could not answer a question got sent off to find the answer while everyone else waited. In any case, unpaid overtime was almost mandatory. "Going home early?" people asked if someone left the office at 6:30 p.m. When workers wore the same clothes two days in a row, it generally wasn't because they'd been out having casual sex in the freewheeling style of their catalogue, but because they had been pulling all-nighters at their desks.

"You give up your life, and all self-esteem and respect for yourself," said the former employee. "Because they take it. You're basically just stupid." What you get in return is a learning experience. "Because they run on such short timelines, you learn what factories and fabric suppliers can really do. You learn how you can really run your business, and run it right. That, they do well. And you learn how to treat people, because they treat people totally wrong."

Many employees quit, having had one too many epithets or legal pads hurled at their heads. They quit even though it can mean having to pay back a $6,000 relocation allowance. Employees who marry and have kids also often quit, because the unwritten message is that they are now out of touch with the youth market. All this seems to suit the company fine. As a spokesman once put it, "Our plan is not to grow old with our customer."

Many alpha apes would approve, and some of them are on the payrolls of Fortune 500 companies. Most are careful not to talk out loud about the strategic value of acting like a jerk. They know that people who talk about it out loud, as did "Chainsaw" Al Dunlap, often end up getting hoist on their own chainsaws. They also know that many commonplace forms of bad behavior violate company policy, even though corporate leaders ignore or even tacitly condone bad behavior as long as the bottom line looks good. Company policy notwithstanding, practitioners of the primordial demonic-male style of management often seem to get ahead, while nicer people falter.

For instance, a senior vice president at an American telecommunications company in the late 1980s made a memorable impression on a colleague as "the biggest jerk I ever met in a corporate setting. He liked to begin every interaction with a sneer and an insult—usually a blunt reminder that he was important and you weren't." Even when he had arranged for his assistant to call you into his office, he would still start by asking "Why are you here?"

If you had had the misfortune to initiate the meeting, on the other hand, he would turn his back and attend to some presumably urgent business on his computer while asking, "Do you have any idea how much this meeting is costing the company?" Meaning *his* salary, not the pocket change they were paying you.

"People who worked for him openly joked that meeting agendas had to allow for a few minutes of 'the dominate-me game.' " The vice president for sales once privately commented: "Geez, I wish he'd just pee on me and get it over with."

Though the constant maneuvering for advantage antagonized almost everyone, it did not seem to hurt the jerk's career. He ran an effective organization that delivered results, says the colleague, and that's all the company cared about, so long as the jerky behavior never got seen by outsiders. The senior vice president has since moved on to become CEO of a prominent health care company. Despite a lagging stock market performance, his 2003 compensation totaled $4.2 million, not counting options on 450,000 shares of the company's stock. This is apparently the payback for being mean.

Where do these people come from and why do they act like that? Scientists sometimes use a type of experimental mouse bred with the oxytocin gene knocked out. The so-called knockout mouse suffers, as a result, from social amnesia; it can't form social memories or normal social relationships. It becomes a cipher, focused entirely on food, sex, shelter. It's tempting to think something comparable happened to the many managerial knockout mice in the corporate world—for instance, the executive who says "Nice to meet you" to a subordinate's spouse, seemingly unaware that they have met seventeen times before. (Next time it happens, the transparent spouse should say, "Nice to meet you, too, Mr. Knockout Mouse." He will probably remember you thereafter, though not necessarily in a way that will advance your spouse's career.)

But there are no doubt many different ways to make a bottom-line personality, genetic mutation least among them. The disturbing question is why these people survive and even prosper in the light of day.

Among the plausible theories commonly put forward by unhappy colleagues: Jerks have no friends or outside life, so they channel all their dark energy into getting ahead at work. Jerks put themselves forward for promotions where nicer types hold back. They're shrewd about showing their sunny side to superiors, even while jerking around their subordinates (the *"Strassen Angel, Haus Teufel"* strategy). Or conversely, they realize that the guy at the top is perfectly content to have his lieutenant act like a jerk, so he can pretend to be a nice guy (the mad dog–frat boy partnership). Finally, and most disturbingly, they succeed because they often run effective organizations. Fear apparently motivates more than we care to admit.

RANDOM ACTS OF CRUELTY

Inspiring fear with random acts of unprovoked hostility may even make a kind of strategic sense. (We're taking the devil's advocate point of view, remember.) We expect conflict when there's something precious at stake—

for instance, food, territory, a step up in rank, or a potential sexual partner. And in these predictable circumstances, we also expect conflict to follow the rules of gradual escalation, with the rivals making ritualized displays of strength and aggressive intent: One guy glowers. The other guy glowers back. They size each other up. If necessary, they escalate to the next stage (standing too close) and then the one after that (bumping and shoving), in a carefully choreographed round of signals and countersignals, until one of them says, "Gee, maybe I'd better just let it go." That way, dominance gets settled without anybody having to get hurt.

But it doesn't always happen like that in the real world. Baboons, and certain managerial types, sometimes launch an all-out attack for no good reason whatsoever. It's a placid afternoon on the savanna. Everybody is sitting around grooming or rooting in the grass for odd bits of food. A low-ranking female is politely minding her business. Suddenly an alpha female comes shrieking down on her in an unprovoked, irrational fury. The startled subordinate flies straight up in the air and explodes in an exophthalmic, piloerectory adrenaline scatterburst, then goes bolting into the undergrowth in search of refuge, with her antagonist crashing close behind. What goes on here? And can it help us understand the same sort of behavior in the human workplace?

UCLA anthropologist Joan B. Silk described the advantages of acting like an unpredictable jerk in a paper with the tongue-in-cheek title "Practice Random Acts of Aggression and Senseless Acts of Intimidation: The Logic of Status Contests in Social Groups." Silk argued that, for baboons, "randomly timed attacks on randomly selected subordinates are part of an evolved strategy that has been favored by natural selection" because such attacks "enable aggressors to deliver the maximum damage to their victims at minimum cost to themselves." The surprised subordinate doesn't have time to escape, muster a counterattack, or get allies to rush to her defense, so the risk to the aggressor is small.

But the cost to the victim can be devastating. Punishment is far more stressful for her when it is unpredictable. She never knows when it is finally safe to relax. So her stress hormone levels remain elevated all the time. Over the long term, this tends to suppress her own aggressive behavior. She doesn't even try to resist, and this is what the big jerk apparently wants: rivals who leap to get out of his way and subordinates who do his bidding without question.

HIS HANDWRITING ON YOUR NEURONS

New managers are often particularly ferocious at the start of a relationship; they apparently subscribe to Kipling's idea about the value of whacking a pony on the nose at the outset. And at the most rudimentary biological level, it seems to work, as the victim's own brain conspires to perpetuate the lesson this opening gambit teaches.

A slightly malicious experiment dating back almost a century demonstrated how bad bosses and other disturbing personalities can take up residence in the autonomic nervous system of their victims: Swiss psychologist Édouard Claparède had a patient with brain damage which left her incapable of forming conscious memories. Every time Claparède entered the room he had to introduce himself all over again. He was curious to know what was going on in the patient's mind. Or maybe she just got on his nerves. In any case, one day he came into the room where she was waiting and introduced himself in the usual way, then shook her hand. She immediately snatched her hand back, because Claparède was holding a pin between his fingers. Next time Claparède offered his hand, the patient would not shake it. Though she remembered nothing else, some part of her knew that Claparède was a bit of a prick.

Claparède had stumbled onto evidence that the brain remembers different things by different means. Names and places and last quarter's results from the Cheez Whiz division take what New York University neuroscientist Joseph LeDoux has dubbed "the high road" to the cerebral cortex, the part of the brain concerned with weighty stuff such as language, problem solving, and impulse control. The cortex is what makes us human, having tripled in size over the course of our evolution from some chimplike ancestor. Pinprick handshakes and other occasions of pain, fear, and danger also take the high road, registering in the consciousness—except for people, such as Claparède's patient, who have suffered damage to the cortex.

But in all of us, the signs of fear and danger take the low road first, sending information fast and dirty to a small almond-shaped region of the brain called the amygdala, from the Latin word for "almond." If the cortex is the part of the brain that makes us human, the amygdala is one of those parts that most clearly make us animals. This is the seat of all fear.

In the early twentieth century, when they mistakenly believed that the sensory and motor cortex controlled the fear response, researchers performed a macabre experiment in which they removed the entire cerebral cortex from laboratory cats. To their surprise, the cats still responded to a threat by crouching, arching their backs, retracting their ears, unsheathing their claws, growling, hissing, and biting. The explanation, LeDoux believes, is that their amygdalas were intact, still receiving and sending out signals on the low road.

The amygdala responds automatically to what LeDoux calls "natural triggers," such as snakes, which were significant threats during our early evolution. When you spot a snake, the amygdala instantly fires out the signal to make you freeze in your tracks. This signal kicks off a cascade of other physiological effects. The adrenal glands release stress hormones. Eyebrows rise and eyes widen so you can see the threat more clearly. Blood vessels constrict and the heart pumps harder in anticipation of a quick escape. The bladder and colon prepare to empty. Your entire body is primed for fight or flight. This is about the point at which your plodding conscious mind finally says, "Oh . . . snake!" And if you depended on that for your survival, you'd be dead.

The amygdala also responds to "learned triggers" from daily life. If someone or something has badly scared you, the amygdala makes sure that you not only remember but also instantly respond to the same threat in the future. It's a survival mechanism. The amygdala doesn't store the detailed memories of why this person is such a snake. The learned trigger—the sound of the voice, the shape of the face, the scent of the aftershave—merely starts the fight-or-flight response.

The signal from the amygdala can also cause clusters of neurons in other parts of the brain to fire in tandem, releasing a flood of conscious memories from the cortex and thus increasing the degree and complexity of the automatic response. The connections among these so-called cell assemblies can be extremely difficult to erase. Thus even when career counselors finally convince him of the need to reform, an abusive boss may find for months or years afterward that hackles still rise when he enters a room. He has in effect written the lesson of his bad behavior onto the neurons of his associates.

This is one reason random aggression is such a dangerous tool in the ordinary business world, particularly within groups of people who need to work together. Though he may not realize it, even the jerk suffers the consequences of broken trust. Among baboons, according to Silk, the major drawback of the randomly aggressive strategy is that it becomes "difficult for dominants to interact with subordinates, even when their intentions are peaceful." The subordinates back away, offering fixed fear grins by way of appeasement.

The same tendency to fearful avoidance can also poison the human workplace. For instance, at Parmalat, the Italian dairy conglomerate, chief financial officer Fausto Tonna was known for his raging temper. After the company declared bankruptcy in a vast financial fraud, one Parmalat employee commented: "If a line manager had a choice between making a mistake or calling Tonna to ask for advice, they go ahead and make the mistake."

NEEDING ANOTHER RAT TO BITE

Once randomly aggressive behavior gets started in an organization, it tends to be contagious, rapidly spreading itself because of a built-in mammalian device for relieving stress, called redirected aggression. Stanford physiologist Robert Sapolsky describes it this way: "Numerous psychoendocrine studies show that in a stressful or frustrating circumstance, the magnitude of the subsequent stress-response is decreased if the organism is provided with an outlet for frustration. For example, the [glucocorticoid] secretion triggered by electric shock in a rat is diminished if the rat is provided with a bar of wood to gnaw on, a running wheel, or, as one of the most effective outlets, access to another rat to bite."

Companies generally do not give bad bosses access to another rat to bite. They give them access to their employees, and the aftermath of a stressful situation is all too often highly predictable. Say your boss's boss has subjected him to an act of random aggression ("What are you, fucking stupid?"). Your boss takes it out on you, then you go and yell at your secretary, your secretary snaps at his girlfriend, and so on in a daisy chain of injured feelings. The diabolical fact of our nature as social mammals is that taking it out on somebody else, even if it means kicking the stupid dog, causes our own level of stress hormones to drop. Watch the dog you just kicked: He'll probably go take it out on the stupid cat, or maybe just chew a stick to smithereens.

At one company, a relatively powerless middle manager practices redirected aggression by taking a photograph of her antagonist from the company newsletter, or some other talisman like a signature, and putting it in her freezer. Sometimes she will also cut it into little pieces, and freeze them inside ice cubes, to which she used to add liberal quantities of whiskey before sitting down to dinner, though iced tea suffices now that she is on the wagon. It's roughly equivalent to the rat having a bar of wood to gnaw on.

"Freezing someone in the ice cubes is a drastic measure," she advises, "only for the most serious cases and severe threat of physical or mental harm." Asked if she crushes the ice between her back teeth, she replies, "That would be gauche." A pause. "Sometimes." Then, reassuringly, she adds, "At the moment, no one is in my freezer."

Biology does not dictate redirected aggression or even justify it. It merely helps explain it. Understanding the natural impulse may also help to prevent it. Studies have demonstrated that, over time, even a decent, compassionate second-level manager will tend to adopt the managerial style of an abusive boss. By being conscious of the tendency, a good number two can short-circuit the process and protect his subordinates from ill feelings on high. If he's smart about his own well-being, he'll take it out instead on a punching bag, a treadmill, or even an ice cube. A smart female boss may be more inclined to cope with grief from on high by seeking the "tend-and-befriend" support of close friends.

Either way, it is "far from desirable," Sapolsky notes, "to live in a social group in which the predominant form of stress management consists of avoiding ulcers by giving them."

THE CULT OF FEAR

Good bosses generally know that fear is a bad thing in the workplace. If they haven't figured it out for themselves, they hear it from consultants, over and over. Fear causes absenteeism, low morale, psychological and physical illness, employee turnover, subversive behaviors, and worse. The pony that gets whacked on the nose at the outset of a relationship is often a pony waiting for the right moment to plant a hind hoof in its tormentor's breadbasket.

But in some primitive corner of their brains, many corporate executives, and even in some cases their subordinates, do not believe it. For instance, Intel's Andy Grove writes: "The quality guru W. Edwards Deming advocated stamping out fear in corporations. I have trouble with the simplemindedness of this dictum. The most important role of managers is to create an environment in which people are passionately dedicated to winning in the marketplace. Fear plays a major role in creating and maintaining such passion. Fear of competition, fear of bankruptcy, fear of being wrong and fear of losing can all be powerful motivators."

Corporate leaders thus often encourage a cult of fear, using the language of threat and intimidation in the apparent hope of winding their troops up into a motivational frenzy. It seems like a useful way to rally mere wage slaves into a fighting force. Crushing the enemy, Dartmouth business professor Richard D'Aveni has argued in his book *Hypercompetition,* is a more compelling mission than merely becoming number one. But this sort of motivational frenzy can also seem a lot like the members of a chimpanzee raiding party leaping up and down, hooting, and flinging excrement to prepare themselves psychologically for the assault.

For instance, as president of Pepsi USA in the 1980s, Roger Enrico named his campaign to develop a lemon-lime soda Overlord because, as he put it, "if this worked, 7-Up and Sprite would feel as if they were on the receiving end of the Normandy invasion." Enrico apparently had visions of himself as Sergeant Rock hitting Utah Beach with bandoliers of Slice over both shoulders.

Leslie Wexner of The Limited likewise once launched a sales campaign called "Win at Retail," or WAR. Following a big-screen presentation of actual battle footage, Wexner strode across the stage like Patton and admonished his sales reps that "retail is war!" This at a company built on "fashion for the woman who feels sexy and confident in all aspects of her life," apparently including the aspect that involves rocket-propelled grenades. The company later renamed the sales campaign "Must Win" (pronounced with teeth clenched and a somewhat muted urge to kill), apparently in belated recognition of the potential for cognitive dissonance.

In fact, selling peignoirs isn't war. Business isn't war, and the great danger of getting people revved up in a warlike frenzy is that animal emotions overwhelm the rational ability to parse out behavior in an intelligent way. The chest-thumping bids to "crush" or "kill" or "bury" outside rivals often translate into hostile behavior inside the company, too.

At WBBM, the CBS television affiliate in Chicago, management at one point pitted the different news shows not just against rival stations but also against one another. So if the producer of the six o'clock news got a tip, she would sneak a crew out to get the story. The producer of the five o'clock news would find out by watching the later show not only that he had been scooped but that his coworkers had withheld an important piece of news from him.

The executive producer pitted people against each other both personally and professionally. He would shout out a staffer's name and then abuse him in front of a newsroom full of colleagues: "This is the worst piece of shit I've ever seen. If you're not careful, you'll end up on the weekend newscast." (A two-for-one insult, with the added value of demoralizing any weekend newscaster in earshot.) Sexual relationships between executives and on-camera talent compromised decisions about story assignments.

And yet the cult of fear persists at such organizations because it often seems to work. One veteran of WBBM recalls it as "the best newsroom in the country and the most dysfunctional place on earth. It was mean-spirited and competitive, but everybody had their lives wrapped up in the quality and reputation of the product. At parties, they didn't say, 'I'm a television producer.' They said, 'I'm a producer at WBBM news.' " Says one former producer, "People were cranked up into such a frenzy that they produced an astonishing quantity and quality of work. Because of fear."

Ex-employees of Miramax also commonly remember working there as the most exciting experience of their professional lives, though bosses Bob and Harvey Weinstein made a name as the incarnation of anger and intimidation even in an industry built almost entirely on the spirit of redirected aggression. Dennis Rice, former marketing head at Miramax, told writer Peter Biskind: "People hate working there, but they love what Miramax stands for, they love the magic they created in the independent film world. It's a very intoxicating feeling, and to be associated with that is addictive, so you found a way to turn a blind eye to how they did it, and to the people who became casualties along the way." Jack Foley, a former vice president of distribution at Miramax, remembered working for the Weinsteins this way: "They were nuclear in their energy and in their anger, even in their malevolence, but to be out there with them, that was the best."

What's going on here? Does the cult of fear ever really make sense?

A CASE FOR TYRANTS?

Everybody knows that working for an abusive boss can be deeply painful. There's also plenty of evidence that abusive behavior is both commonplace and costly. In recent studies, the percentage of workers reporting that they had been bullied over the previous year ranged from a low of 4 percent among steelworkers to a high of 38 percent among employees of Britain's National Health Service. (The wide disparity may reflect differences in the definition of bullying, which can encompass anything from physical abuse to glaring, according to Joel H. Neuman, who studies the issue at the State University of New York in New Paltz. Industries and nations also vary in their sensitivity to the issue of bullying. Finally, people are no doubt more cautious about bullying steelworkers, say, than pathology technicians.) At any given moment, an estimated twenty percent of people find themselves working for a brutal boss.

Hardly any of them report the experience as "magic," "intoxicating," or "the best." These may be words that occur, if at all, only afterward, once the beleaguered employee has made a safe escape. In one study, roughly half the victims of workplace abuse lost work time worrying about the treatment they received and avoiding their antagonist. Twelve percent actually changed jobs.

The hidden medical costs make the economic and moral case against bullying even more damning. Actual physical abuse is rare, says Neuman. "But the other stuff . . ." He laughs ruefully. "It's a drip-drip-drip water torture. If people suggest you are stupid, or don't give you meaningful work, they don't see you bleeding. They don't see the high blood pressure. But who knows how many supervisors are responsible for strokes?"

The diabolical beauty of random aggression, from the bad boss's perspective, is that these costs seldom get debited against his account or reputation. It may take years for the symptoms to manifest themselves in the form of high blood pressure, reduced immune function, lower levels of HDL ("good") cholesterol, impaired cardiovascular response, atherosclerosis, fertility problems, and other ailments. But the cost to the company is real, in medical expenses and in lost productivity from a demoralized and physically debilitated workforce.

It is unfortunately much more difficult to measure whether the cult of fear might also sometimes improve productivity. But to continue playing devil's advocate for a little longer: Would Intel have thrived under an Andy Grove who wasn't paranoid? Would Miramax ever have become the most important and influential independent movie producer of the 1990s if Harvey Weinstein had been a nice guy? This is a bit like asking what life would be like if a hyena became a vegan. But it is at least worth considering a few plausible reasons why bad—or let's say difficult—behavior may sometimes work, at least in the short term, to the profit of almost everyone.

PREDATORS MAKE FOR HEALTHY HERDS

Contrary to our expectations, it can be demoralizing to work with people who are too nice. When Lew Platt became CEO in the 1990s, for example, his priority was to make Hewlett-Packard a wonderful place to work. He was so focused on helping employees feel valued and secure in their jobs that it began to resemble the civil service.

"People could get dismal performance ratings for four or five years in a row," one HP executive later complained to business journalist George Anders. "It was almost as difficult to get rid of them as it was to get rid of a tenured teacher. People felt they were owed lifetime employment." Platt believed, reasonably enough, that it was destructive to behave too harshly with his own employees. But the natural world suggests that this is true only up to a point.

Animal communities typically depend on the discipline imposed by hungry predators hankering for their throats. They certainly do not like it, and they suffer anxiety over the short-term prospect of being weeded out as one of the weaker members of the herd. But a moderate sense of threat is healthy; it keeps them alert, with one eye always on the lookout. Over the long term, the herd becomes stronger through the painful process of having its less nimble, less vigilant members ranked and yanked by predators. Lose the predators, and everybody gets fat and lazy.

In one recent study, for instance, scientists looked at a population of guppies taken from a high-predation area and transferred to an area with little or no predation. The sudden absence of natural selection caused "a striking, evolutionary loss of escape ability" over little more than fifteen years, or about thirty guppy generations. Being vigilant is evidently a costly behavior, so the guppies gained in the short term by the unexpected opportunity to relax. But they ultimately became weaker and more vulnerable. Likewise, employees at Hewlett-Packard undoubtedly enjoyed the short-term sense of security Platt afforded them. But in the long term, the company went flat.

It is of course misleading to characterize corporate leaders as predators within their own companies. They're more like shepherds continually culling weaker animals from the herd and sending them to the, um, outplacement office. They're like herding dogs, simultaneously protecting their employees while also nipping at their ankles to keep them moving in the right direction.

And Platt was an ineffectual sheepdog. In his book *Perfect Enough,*

George Anders described Platt's failure to produce change during a managers conference in Monterey, California: "Six years of nurturing the softer side of HP had left employees and managers unwilling to take any sudden steps until everyone had talked things over and decided as a group that they liked the new direction. 'Terminal niceness' had set in. In that environment, Platt might as well have tried to shoo seagulls from the coastal rocks outside the conference center. He could create a momentary flutter, but he couldn't really change anything."

THE STRANGE ALLURE OF THE MEAN BOSS

People often prefer, counterintuitively, to hang around with individuals who take advantage of them. This may seem at first like one of the sadder quirks of human nature. Researchers who studied Machiavellian behavior in the 1970s and 1980s were often dismayed to find their study subjects strongly attracted to manipulative, exploitative individuals. In one case, the researchers themselves admitted with chagrin to experiencing "perverse admiration" for these so-called high-Machs.

At first they thought the high-Machs were merely adept at "impression management"—that is, they were charming or clever enough to get what they wanted without being too obvious about it. So the researchers constructed written tests to screen out the influence of face-to-face attractiveness. In one study, for instance, people were asked to write about what would happen if they got stranded on a desert island with two other individuals of the same sex. The low-Mach types wrote warm tales about sharing limited supplies. One high-Mach female, on the other hand, wrote: "Mary and Jane are cold bitches who constantly complain . . . when I got really hungry I wondered how I could cook them with the limited cooking equipment we had."

Test subjects understandably rated the characters in the high-Mach stories as "more selfish, uncaring, judgmental, overbearing, untrustworthy, aggressive, undependable, and suspicious." The test subjects didn't want to share an apartment with them, much less a desert island, according to biologist and game theorist David Sloan Wilson. But they accepted high-Mach characters for "relationships that involved working as a group to manipulate others"—that is, for relationships like the ones we form in the business world.

Test subjects didn't necessarily admire bullies. They were attracted to schemers and connivers, people who were clever enough to take advantage of them without being too brutal about it. Sociologists typically draw a line between bullies who use coercive dominance and the more adept or "prosocial" individuals who find nicer ways to get what they want.

But it can be a fine line. Bosses in the human workplace routinely shift their dominance style from prosocial to coercive and back again, depending on the audience, the circumstances, or what they drank at lunch. A prosocial boss, the decent and likable head of a hospital emergency room, may bark out orders or rebuke an inept orderly when the victims of a five-car pileup come wheeling in on a Saturday night. The sales manager at a software company may become hostile and demanding as the end of the quarter approaches, and people often respond by working harder.

In a recent study of adolescents, Patricia Hawley of the University of Kansas categorized dominant individuals not just as coercive or prosocial but also as bi-strategic. The bi-strategics used prosocial techniques such as offering unsolicited help, reciprocating favors, and building alliances. When it suited them, they also employed classic bullying techniques including threats, social exclusion, destructive gossip, and physical aggression. Not surprisingly, peers and subordinates ranked the prosocial types as the most attractive dominants. But the bi-strategic sorts also proved surprisingly popular, and far more attractive than any plodding subordinate, even though survey participants seemed entirely aware of when they were being coerced or manipulated.

The "allure of the mean friend," or the mean boss, isn't simply masochism. Nor is it just a product of our evolutionary predisposition to follow dominant individuals. What bi-strategics bring to the relationship, Hawley suggests, are genuine social skills; they are exciting to be around. "These kids who are supposed to be repellent aren't. They make good friends (unless you get on their bad side)." No one has studied the material advantages of hanging around with a manipulative, exploitative, or even occasionally abusive individual. But the benefits may perhaps at times outweigh the costs. Particularly in business, it can pay to work with someone who will do what it takes to bring home the booty.

Even in terms of our own individual performance, we often do better work for a more critical and demanding boss. It is, of course, possible to motivate people without abusing them. But there are also times when we seem to accept abuse as appropriate to the circumstances. When the heavyweight eight is closing in on the final five hundred meters of a crew race, for instance, the otherwise gentle little coxswain will loudly question the man- or womanhood of every rower and even bawl out individual insults, and you can feel the boat surge with the renewed effort. A coxswain who keeps it gentle generally does not get invited to the next race.

Likewise, what the ex-employees of WBBM and Miramax are probably remembering is not the pain of having been bullied and browbeaten but the peculiar satisfaction of having achieved things that seemed to be beyond their own limitations. Because they were driven to it. Because of fear.

One time in Kenya, a primatologist was traveling in a remote area when his vehicle broke down. He had no choice but to walk several hours across grassy savanna before nightfall. Unfortunately, a pride of lions found him before he reached camp. He grabbed the nearest tree and scooted up out of reach, remaining there until the lions grew bored and wandered off. Later, when he went back for his vehicle, he stopped to look at the tree. It was little more than a stick, devoid of handy branches or toeholds. There was no way on earth he could have climbed it, he realized, without a lion at his heels.

TIME TO BEHAVE

But enough. Let's take off the devil's advocate mask and come back to reality. Bosses who rule by fear may win battles but often lose the war, driving their companies into decline or even bankruptcy, as happened at Sunbeam, Enron, and Kmart. Individuals who affect the confrontational style too openly also frequently come to unfortunate ends, or at least to unexpected rest stops in midcareer. As SUNY researcher Joel Neuman puts it, with scholarly precision: "That shit catches up with people."

▪ When Durk Jager took over as CEO at Procter & Gamble in 1999, he publicly denounced the company's tendency to "Procter-ize employees," making them sound alike, think alike, and look alike. What he wanted, apparently, was to Jager-ize them. When the *Wall Street Journal* asked if he was as mean as reputed, he smiled and said, "A lot worse." Jager, a veteran of the Dutch air force, joked that Dutch lessons would soon become company policy. Wall Street analysts praised Jager as a "gifted strategic thinker" with a knack for finding the heart of a problem and putting his scalpel there. But Jager's attempts to shake up the company were so sudden and so blunt that he created global turmoil. By the time the board dumped him after just seventeen months on the job, P&G stock was down 37 percent.

▪ John J. Mack, known as "Mack the Knife" for his cost-cutting savvy and also for his imperious style, lost a brutal battle to run Morgan Stanley. He soon moved on to Credit Suisse Group, which he quickly restored to profitability. But his abrasive ways clearly rankled. At one point, Mack was asking the Zurich-based board of directors to increase compensation for his New York banking executives. When directors questioned his motives, he snapped, "I'm not doing this for the money. I don't need money. I'm wealthier than any of you. What I need is the money to keep our people in place."

He was probably right. Soon after, though, the board unceremoniously dumped Mack from his job as co-CEO. The company cited strategic differences. But three separate sources mentioned the "wealthier than you" remark to the *Wall Street Journal,* suggesting that Mack's crude assertion of dominance stuck, in a very personal way, in the craw of his more discreet European colleagues.

▪ Sanjay Kumar, son of an animal behavior researcher in Sri Lanka, built an empire at Computer Associates on what he conceived to be a "Darwinian" model, fostering a culture of extreme challenges, sharp disagreements, and abrupt dismissals for underperformers. Even customers sometimes got ground down in the process. When Albertson's resisted pressure to sign a multiyear license in 1999,

Kumar and his associates threatened to shut down the existing software on which the entire grocery chain depended. In a good year, revenues at Computer Associates jumped by 33 percent. But ultimately the company became a synonym for rancor, mistrust, and bad business practices. An accounting scandal forced Kumar and other top managers to resign, prosecutors have filed criminal charges, and investors are suing to recover more than $1 billion in executive bonuses based on false accounting.

The opportunities lost through abrasive, unpleasant behavior are seldom so public. People simply withhold things. It happens every day, in a hundred thousand small ways, so the guilty party is never even aware of it. Sometimes the loss is trivial: At a hip music store in Red Bank, New Jersey, not long ago, for instance, a surly customer ordered the counter clerk to look up a certain artist on his computer. What the harried clerk typed on the title search line was *"Go away, leave me alone, die, die, die!"* Then he put on a regretful expression and said, "I'm sorry, sir, we don't seem to have that in stock today."

Other times, the loss can be incalculably large, as when subordinates withhold dissenting opinions from an overbearing boss. During the last flight of the *Columbia* space shuttle in February 2003, for instance, NASA staffers failed to voice their real concerns about the damaged spacecraft because they feared "they would be singled out for possible ridicule by their peers and managers," who had an overwhelming drive to get on with business as usual. Management had somehow allowed this closed-minded culture to persist despite NASA's own tragic record of having suppressed dissent before the 1986 *Challenger* space shuttle disaster. In the two accidents, fourteen astronauts died.

Creating a cult of fear can also lead workers to retaliate. At a New Jersey chemical company, management laid off fifty workers. One of them, an executive with a $186,000 salary and thirty years on the job, deeply resented the use of guards "to escort us off the premises like criminals." He went home to commit an act of computer sabotage, which cost the company $20 million.

Abusive leadership might perhaps work, or at least *seem* to work, at a Miramax or a WBBM because people are engaged in relatively short-term creative projects. So it's easy for them "to identify with the creative group and to put up with lousy leadership," suggests Terry Pearce, an executive coach. "And then, if the product is good, it's something you were part of that you can point to for the rest of your life." But most people's working lives aren't like that. "If you're in, say, financial services, or you make chips for a living, it's really hard to identify with a specific project that you would be excited about getting out the door and having it be complete." In the day-to-day grind of ordinary work, with minimal opportunity for creative expression, people need a boss who appreciates their work, not a bully or a jerk. "It's tough if you make ties or sell hamburgers."

Pearce specializes in coaching corporate chieftains to mute their antagonistic and overbearing tendencies. He starts with a 360-degree survey and a round of interviews with subordinates, then lets the boss hear anonymously what the people around him really think of his management style. This can be a shock. Most leaders never recognize how domineering they can be, "they just do it."

Then Pearce works to help the boss understand how this behavior distorts the organization. You can use fear to drive people, Pearce says, if you want an organization "filled with people who do what they're told, who only bring 75 percent of themselves to work," who give their compliance but not their commitment, who refrain from offering their creative ideas, and who almost always have their resumés out on the marketplace, meaning a high rate of turnover.

Most clients recognize that this isn't the kind of organization they want to lead. Or at least it's not the kind their board of directors wants them to lead. And they begin the painful process of change. But not always. One client listened to a devastating review of his Chainsaw Al managerial style and then said, "Do I have to change?"

"Not unless you want to stay here," Pearce's partner replied.

"Well, that's fair," the client replied.

"The style that you operate with gets instant short-term results. But people aren't going to work for you very long," the partner explained. "If you want to be a turnaround specialist for the rest of your life and you don't want to lead anybody, then, absolutely, you can do that."

Says Pearce: "And that's what he did. He left the company two or three months later. He's a turnaround guy. He goes in for two or three months, gets it right, cuts the heads, leaves town. Is he a success? He's rich. People pay him well to do their dirty work. But he's not a leader."

IF A BABOON CAN DO IT . . .

Learning the opposite managerial style can be a struggle. An individual who may have gotten to the top by being ruthlessly aggressive suddenly discovers that he needs to listen, to learn, to show pain, to connect with people on an emotional level (other than terror), and to instill trust. It's like a Chicago alderman waking up one morning and discovering he has to act like Jimmy Carter. It's like an Afghan warlord trying to be Kofi Annan.

It may not be much consolation, but even surly, belligerent baboons apparently do not much like bullies, and get along just fine without them when they can. And even among baboons, the culture of a group can apparently induce good behavior in alphas who might otherwise be absolute bastards. Robert Sapolsky and Lisa J. Share, both Stanford physiologists, have been studying savanna baboons in Kenya's Masai Mara Reserve for more than twenty-five years now. In the mid-1980s, all of the biggest, meanest males in the Forest Troop there ate meat tainted with bovine tuberculosis at a tourist lodge garbage pit and got wiped out by the disease. The females and the more mild-mannered males survived because they did not dare to visit the dump.

The researchers stopped watching Forest Troop for a while. But when they came back in 1993, they found to their amazement that it had become that rare thing in baboonology, a relaxed, supportive, and even gregarious culture. The dominant males spent an unusually large part of the day hanging around in close proximity with adult females, infants, adolescents, and juveniles. Instead of thrashing and biting to get their way, they seemed to rely more on affection and mutual grooming.

This isn't to say they had suddenly become nice. "We're still talking about baboons here," says Sapolsky. (He once summed up baboon social life this way: "If baboons are spending only four hours a day filling their stomachs, that leaves them with eight hours a day to be vile to one another.") But they became more selective—one is tempted to say more rational—about when to be vile. In most baboon troops, dominant males spend much of their time approaching and displacing low-ranking subordinates. It is pure harassment, since the low-rankers pose no competitive threat. The constant bullying causes low-rankers to suffer chronically elevated levels of the stress hormone cortisol.

In the new, improved Forest Troop, the dominant males also engaged in approach-displacement behavior. But it was almost entirely directed at the close-ranking males who were their true rivals. Not only did the low-ranking subordinates escape the myriad health problems associated with chronic stress, but even when the researchers injected them with a drug that produces a short-lived state of anxiety, they remained visibly more calm. In human terms, this suggests that unbullied subordinates would be more productive and capable of responding better to real threats.

For Sapolsky and Share, the most remarkable thing about the reformed Forest Troop was that none of these benignly dominant males was a mild-mannered survivor from pre-tuberculosis days. When baboon males become adolescents, they typically abandon the troop in which they were born and go out to seek a place in a neighboring troop. So all of the enlightened males were actually newcomers. Something special about Forest Troop had induced them not to follow the usual baboon fast track and become mean bastards.

Sapolsky and Share noted that the females, who are the keepers of baboon culture, treated the newcomers in the same affiliative manner they used on resident males, even though the newcomers often started out acting vile toward them. The "social capital" of living in a healthier, more supportive community apparently enabled the females to put up with the abuse until the "jerky new guys" gradually figured out that "we don't do things like that around here."

To put it in human terms, it's like the executive assistant who, confident in her own abilities and in the support of the workplace community, gently lets the belligerent CEO know that "it's just not possible for us to work together like that." It's like the nurses in one hospital who respond to what they call "Code Pink," when a doctor loudly berates a nurse. The nurses walk over and stand beside the object of wrath. Then they stare at the doctor in silent puzzlement.

Most doctors (like Forest Troop baboons) eventually figure out that public displays of nastiness and aggression are probably not, after all, the best way to get what you want in a civilized society.

When humans wear epaulets, it's to make themselves look bigger and more powerful than they really are. The redwing blackbird wears epaulets for the opposite reason, as a way of looking harmless. The male is entirely capable of putting on a fierce, militaristic show. He has bright red shoulder patches, which he uses when necessary to advertise that he controls a territory. But this badge of status can also elicit violent attack from rival males.

So redwing blackbirds have evolved an epaulet, which fans down over the red shoulder patch and conceals it completely. This allows them to scout around in plainclothes when they are out prospecting for a better territory, or sneaking around in a neighboring territory to consort with another male's harem of females.

Even on their home territory, the males seem to recognize the value of playing down their status. When researchers clipped the epaulet covers of some birds as part of an experiment, males thus obliged to go around "permanently signaling combativeness" had to spend more time fending off intrusions by neighboring males. These males got treated "as if they had broken the contract of neighborliness by signaling too much combativeness and were therefore no longer accepted as partners for peaceful sharing of boundary zones."

Sometimes it simply does not pay to act too fierce.

15

RUNNING WITH THE PACK
Why Lone Wolves Are Losers

It's an ant colony. A new business opportunity drops on the ground, 5,000 people swarm on it, and when it's gone, they all disappear to some other project.

—An Intel employee on his company's culture

So why are so many people unhappy in the workplace? Why are they ineffective?

Partly it's because many employees don't know what they're doing there. They have no sense of where their work fits in the larger scheme.

Worse than that, they often don't know the people with whom they are supposed to be doing it.

And both problems arise not just from the sprawling scale of modern life but from the assumption that we can routinely disregard our fundamental nature as social and emotional animals and get away with it.

WHO THE HELL ARE YOU?

In the natural world, most social animals spend their lives in one group and quickly figure out who everybody else is and where they fit in the picture. Even sheep can identify fifty different faces (sheep faces, that is). But modern corporate humans change jobs so often, and put in so much travel even within the same job, that they often cannot name the person in the next cubicle, or sitting across from them in the conference room. They may not get their kids' names or their stepkids' names straight on their occasional sorties home.

The sense of being just an interchangeable cog in a big machine has been a fact of blue-collar life at least since Frederick Taylor brought his time-and-efficiency methods to the assembly line early in the twentieth century. But the disorienting sense of anonymity now afflicts even top executives, who may have whole divisions reporting to them from locations around the world. An honest managerial job description these days might read something like this: *The candidate will parachute from the tops of tall buildings, make sense of what is happening on each floor as he or she plummets past, and implement cost-effective methods to make it happen more efficiently before reaching the ground. Repack parachute, move to next tall building, repeat.*

When Michael Capellas became CEO at WorldCom, the bankrupt telecommunications company (since renamed MCI and tentatively scheduled to become part of Verizon), he held a meeting with his top one hundred executives and asked who could identify everybody else in the room. Nobody raised a hand, not even when he asked if they could name just half the people in the room. Capellas could have said, "Come on, people, even sheep can do this." Instead, he decided to concentrate top management in his Virginia headquarters, so executives could rub shoulders and thereby feel greater accountability and the visceral need not to let down their teammates, traits notably lacking at the old WorldCom.

Capellas was on the right track. No human organization can succeed, or even survive for long, unless it acknowledges that its employees are social animals. This doesn't necessarily mean putting them in the same location. But every organization needs to find some means, in the shift and sprawl of modern life, to gratify its employees' powerful instinct for affinity. Humans evolved to spend much of their lives in the same family, the same clan, the same tribe, the same landscape, among familiar faces bound together by a history of shared experiences and often by ties of blood. It was routine to know and trust the person working at one's shoulder, and this bond was sometimes a matter of life or death. Everything in our fundamental makeup is still geared to function best in these small, close-knit groups.

In the modern world, the job is often the main place we get to fulfill this social craving, because other, more traditional forms of community have largely vanished. We often spend more of our waking hours at work than we do at home. Romances, friendships, and roughly 40 percent of marriages begin at work. Yet our relationship with the workplace community is often necessarily tentative. Companies often claim to be "just one big happy family," but everybody knows they make a poor substitute for the real thing. In the past, a company could hold out the promise of lifetime employment and cradle-to-grave benefits and thus elicit genuine, almost familial feelings of allegiance from employees.

But relationships get heavily discounted in a global economy where the main question companies and customers alike have in mind is, "Can I get it cheaper somewhere else?" In most big happy families, Mom and Dad try not to say things like, "Timmy, if you won't eat your vegetables, we can offshore that work to a starving child in China."

BIG IS DIFFICULT, NOT IMPOSSIBLE

Getting to know fellow workers or developing a relationship with them, let alone a sense of mutual obligation, can seem impractical when businesses sprawl across scores of nations and employ hordes of people who may never even meet. An individual built by evolution to live face-to-face in a group of perhaps 150 people can easily get lost in a corporation of up to ten thousand times that size. Hence the modern dread of events spinning beyond our control.

Our salvation, even in the most far-flung organization, lies within the group. In large crowds, we are little more than numbers. But in small groups, even small groups within vast corporations, we become human. "We can get people to feel positive about the corporate brand," says London Business School professor Nigel Nicholson, "but people identify first and foremost with the small group they belong to. Only through such groups can you usually get people to make sacrifices for the business as a whole."

In World War II, Easy Company of the 506th Regiment, 101st Airborne Division started out as just another random assemblage of young men. But the bond these men formed during almost two years of preparation, most of it under the modest, competent, self-sacrificing leadership of their officer, Richard Winters, kept the company fighting despite 150 percent casualties. Individuals became subsumed in the group. Letting one another down became unthinkable. The strength of this bond was what made the book and television series *Band of Brothers* so powerful—and, not incidentally, helped to defeat Hitler. "Were you a hero?" a grandchild later asked an Easy Company veteran, who replied, "No, but I served in a company of heroes."

Even in the humdrum world of daily work, the social bond between coworkers is what makes a group succeed. In the best groups, it's what enables people to do more and become better than they ever dreamed possible. The bond forms spontaneously, not usually from the top down, but laterally, largely through casual chat and unscripted time together. Smart managers recognize that these social interactions among their subordinates can be the chief instrument of their success. Weak managers often regard these interactions with dismay and may even attempt to obstruct them.

For example, at NYNEX, a regional telephone company, the system for delivering high-speed T1 transmission lines in the early 1990s was going bust. Without consulting the people on the ground, management had installed a computerized system to automate the work and make it more rational and efficient. But the process was taking thirty-five days from the time an order was placed to the actual installation of the T1 line. The company (now part of Verizon) was disappointing important customers and losing market share.

Management thought the solution was a new computer system. But when a corporate anthropologist named Patricia Sachs studied the problem, she found a typical case of a company paying attention to time and cost questions and no attention at all to how people really get their work done. The company's "trouble ticketing system" was essentially a high-tech version of old-fashioned Taylorism. It broke the work down to its logical parts, digitized them, and treated individual workers as little more than input/output mechanisms.

Workers told Sachs that in the past when they had a problem, they'd get on the phone and say, "Hey, Jack, listen, I got this problem, can you help me out?" And the two of them would talk for a few minutes and figure it out. To management, this looked suspiciously like schmoozing. The high-tech system was designed specifically to eliminate such conversations. When a worker ran into a problem, he now filed a "trouble ticket," to which another anonymous worker might reply electronically a few hours later. The conversation was eliminated. But "the story line of the problem, if you will," was also lost, according to Sachs, and "so was the work community" because no one knew who else was working on the job. Nobody could be bothered training newcomers either, because the system gave them no credit for that.

The bottom line was that doing the job with the automated process took more time, not less. The solution was to rebuild the old informal social network for troubleshooting problems. Among other things, this meant putting the salespeople and the engineers in the same office, a move that raised hackles at first because union members would be working side by side with management. To satisfy conventional notions of hierarchy, the two groups ended up in adjacent rooms. But they were at least now close enough to call out to one another.

This made it easier for them to coordinate the elements of a job and helped them coalesce into a team. The solution also included hiring a "turf coordinator" to make sure people in the field were getting together with the right colleagues to solve a particular problem. Disaffected workers regained their sense of purpose and group identity. And customers got their T1 lines just three days after placing the order.

THROW MOMMA AT THE TRAIN?

Does getting to know the people you work with really make a difference? Is it worth bothering about, for instance, when the nature of the job means the connection will be brief (as you plunge past with parachute deployed)? Do we form a social bond even when the motives that bring people together are strictly expedient?

Maybe we should put this in terms a manager can readily grasp. At the Center for the Study of Brain, Mind, and Behavior at Princeton University, neuroscientists like to present test subjects with what they call the "trolley dilemma." In one version, there's a runaway trolley, and if it keeps going straight, it will kill five people. But if you flip the switch up ahead, the trolley will divert onto another track and kill just one person. Do you flip it? Most test subjects automatically answer yes.

But let's say this time you're on a footbridge overhead. There's no switch in the tracks, just five people up ahead who are doomed unless you can stop the runaway trolley. Standing beside you on the bridge is a plus-size person who could stop the trolley with sheer corpulence and save five innocent lives. It's clear she's not about to do the right thing by stepping off the edge to a heroic death. So is it okay for you to push her? Almost all test subjects instantly answer no.

Philosophers have often assumed that we make moral judgments like this based on intellectual reasoning. But reason says the consequence is much the same in either case: You sacrifice one life to save five. So what's the difference?

The Princeton neuroscientists used magnetic resonance imaging to study brain activity as test subjects responded to the two alternative situations. When the victims are just statistics somewhere down the track, as in the first situation, the areas of the brain associated with social intelligence and emotion stay quiet and we seem to look at the problem rationally. But our emotions kick in when the connection is personal. These same areas of the brain (the posterior cingulate gyrus, the superior temporal sulcus, and the medial frontal gyrus) were far more active in the footbridge situation. Lead scientist Josh Greene speculates that our long evolution in small social groups may have left us biologically prepared "to be interested in people with whom we interact in a personal way, not in statistical information."

Why does it matter? The trolley dilemma gives new importance to the notion of face time, because it suggests how the emotional connection affects every aspect of our working lives. If a miner gets stuck in a shaft, says Greene, the mining company "may spend millions to get him out." But the same company might have balked at spending pennies on simple safety precautions to prevent the accident from occurring in the first place. The moral is that, whatever you want done, you should put a face on it.

More than that, you should show your own face around the office.

You should chat (but not dawdle). You should network (but not too purposefully). You should be careful not to become too complacent about your telecommuting. The NYNEX example suggests that making the personal connection is often a better way to get the job done. And, incidentally, when it's time to cut the budget, or lay off another three hundred workers, or offshore a few more engineering jobs, you won't just be a statistic somewhere down the tracks. You'll be a face the decision makers will have to push off the bridge with their own bloody hands.

But, oh dear, we were trying once again to talk about cooperative behaviors.

THE VIRTUAL HONEYMOON

Apart from arranging to work in close proximity and allowing for face time, groups seem to achieve their best results when they manage to remain small. Consultant and psychologist Richard Hackman, who has made the study of teams his specialty, has come to depend on "a rule of thumb that I relentlessly enforce for student project groups in my Harvard courses: A team cannot have more than six members." Adding one or two additional members may seem insignificant, but every extra person vastly increases the number of one-on-one relationships, and thus the number of demands on our emotional and intellectual attention. "Even a six-person team has fifteen pairs among members," says Hackman, "but a seven-person team has twenty-one, and the difference in how well groups of the two sizes operate is noticeable."

When a group grows to twenty people, University of Liverpool anthropologist Robin Dunbar points out, "I have to keep track of nineteen relationships between myself and fellow group members, and 171 third-party relationships involving the other nineteen members of the group." The mind boggles. Dunbar uses anatomical considerations to arrive at the same ideal team size as Hackman. He notes that speech discrimination falls off at a fixed rate as the distance between speaker and listener increases. This suggests that, "under minimal noise conditions, there is an absolute limit of about five feet beyond which the listener simply cannot hear enough of what the speaker is saying . . . Even allowing a minimal shoulder-to-shoulder distance of six inches, a circle five foot in diameter would place an upper limit of about seven on the number of people who can hear what a speaker is saying." (For the executive who thought round conference tables served mainly to let everyone see the kill, perhaps it is also a matter of being able to hear the rending of flesh?) Other studies also generally arrive at an ideal size of six or eight people for effective teams, up to and including corporate boards of directors.

So why do workplace teams often balloon to a much larger size? Not with the idea of making them more effective, according to Hackman, but to spread around the responsibility and the accountability (that is, the blame), or to be politically correct by representing all constituencies. It's the herd instinct. He warns that such teams often prove "incapable of generating an outcome that meets even minimum standards of acceptability, let alone one that shows signs of originality." But the managerial spirit is often timid, and larger groups abound. The "Who the hell are you?" question thus routinely arises. And the polite fiction of familiarity means that it usually goes unacknowledged.

In the animal world, first meetings of unacquainted individuals are often tense. Levels of the stress hormone cortisol surge, according to conventional thinking, and strangers duke it out until they establish a social hierarchy. But some biologists have lately argued for an alternative interpretation, that apes and monkeys go out of their way to minimize conflict on first meeting. They avoid large gestures and loud vocalizations. Acts of aggression tend to be ritualized.

"In fact, serious fights are apparently uncommon during initial encounters, and agonistic overtures are rarely responded to in kind," says Sally P. Mendoza, a primatologist at the University of California at Davis. Studying squirrel monkeys, a highly social South American species, Mendoza found that females coming together for the first time sometimes experience *reduced* cortisol levels, which may persist "for several months following the formation of new groups." She argues that different species have different rules, conventions, and perhaps even biological adaptations to encourage the formation of social groups and ease tension as social roles get sorted out.

The "honeymoon phase" during which participants in human groups sit back, speak softly, and defer to one another may be a basic primate behavior. Smart managers should find ways to ease the natural process of familiarization so that every meeting does not feel as tentative as a first meeting. The faster a group achieves a level of trust and comfort, the sooner it can begin to grapple with real issues.

To help people find their place in the constantly shifting assortment of workplace faces, for instance, rewritable nameplates ought to be a standard feature in every meeting room, even if just to say "Tom—Data Processing." It may seem gimmicky, but introductory devices can also help, like going around the room and asking each person to describe, in a sentence, a professional or personal triumph.

Such shorthand techniques can sometimes work even when groups meet only virtually. Ann Majchrzak, a University of Southern California business professor, led a survey of such groups. She reports that participants in a Unilever Latin America team, who were spread out over five countries, actually introduced themselves in early teleconference sessions with a quick reference to their Myers-Briggs personality dimension: "I'm Estefan, and as you know, I tend to think out loud."

The group coalesced and even developed its own language, Portuñol, part Portuguesa, part Español, to accomplish its business.

At Shell Chemicals, according to Majchrzak, a financial work group set up an intranet "team room" with photos arranged in a clock configuration. Group members working at their desks kept the team room open on their computer screens, and speakers could introduce themselves as, say, "Kate at ten o'clock" or "Ibrahim at three." The team room also gave participants access to working documents and minutes of past meetings. They could post biographical notes, including accomplishments, areas of expertise, and interests, giving them a chance to flesh out their profiles a bit. (But carefully! It is possible to cross the fine line into the land of too much information, as when John Rolfe, a former investment banking associate at Donaldson, Lufkin & Jenrette, confessed: "I dreamed that one of my managing directors, dressed in a black leather dominatrix outfit, was flagellating me with a whip made of Twizzlers.")

It may seem retrograde in a virtual universe, and across twenty-four time zones, but telephone contact also remains essential in bringing a group together. Vocal tones convey nuances and emotions that may not be accessible by any other means. People talking by phone listen to the actual words that are spoken, of course, but even more to the spaces between the words. They are quick to pick up the note of genuine enthusiasm, or the perfunctory delivery of a sham emotion.

Face-to-face social gatherings—coffee and bagels in the morning, a beer after work—are still among the best ways to induce people to cooperate on subsequent projects. But this is often impractical in the modern world. The Majchrzak study described a work team at a chemical processing company, involving an unwieldy assortment of forty specialists at locations around the world. When the group had accomplished its goal, with a savings to the company of more than $2 million a year (and without ever actually having met in the flesh), the team leader held a celebratory conference call with a cake delivered to each site. The party was virtual. But the cake at least was real.

THE GROUP

Whether or not anybody intends it, people in the workplace naturally form themselves into groups. We cluster together by department or professional specialty, by ethnic group or gender, or by some outside interest such as ultimate Frisbee or choral singing. We form cliques of like-minded, like-ranked, like-salaried people. Separating ourselves from outsiders and associating with people we resemble, even if only superficially, is a quick-and-dirty tool for re-creating the tight social network of the tribal clan or the chimpanzee troop. It's one way we prepare the ground for the development of trust, cooperation, and reciprocity.

One of the conventions people routinely use to help the members of a group come together is to minimize or remove things that obviously set them apart. It's a way to communicate the feeling that we are in this together, that we're on the same side. The white-shirt dress code at IBM, for instance, got started as a way of making conservative white-shirt clients more comfortable. It vanished when casually dressed customers began to regard it as stuffy and out of date.

At Coca-Cola, similarly, employees obeyed an old company dictum against setting themselves apart: "If your bottler drives a Cadillac, you drive a Buick. If your bottler drives a Buick, you drive a Ford. If your bottler drives a Ford, you walk." Executive Steve Heyer crashed and burned at Coke partly because he seemed to fly in the face of this understated culture. On joining the company, he didn't just move to an upscale neighborhood but bought a house on the very street where cola god Robert Woodruff once lived. And he drove a Mercedes.

Conformity in the way we dress or talk creates a sense of ourselves as a cohesive group. Sometimes the conformity comes in a single ritualistic moment, as when the members of a sports team all shave their heads and stand together as uniform as billiard balls in a rack. More often, the code evolves gradually and seems almost to creep up on the group, the way husbands and wives come to look like each other, or like their family dog, without anyone ever intending it.

At Enron, for instance, the trading floor was the habitat of one such elite group. Traders were "moody, uninterruptible, and heavily bonded with one another," according to Brian Cruver, who worked on the floor briefly before the company's demise. "The first thing I noticed about Enron traders is that they all looked very similar. A goatee was fairly common; otherwise they maintained a clean-cut yet outdoorsy look; and if they didn't wear some version of a blue shirt every day, then it was like they weren't on the team."

Finding himself surrounded, on his first day of work, by a dozen traders in identical blue shirts, Cruver didn't merely notice the culture to which members of the group unconsciously conformed. He also committed the faux pas of referring to it out loud. "When did they hand those out?" he asked the blue shirts.

No one laughed.

Groups also frequently develop their own argot or language (Portuñol, for instance). It's often a practical tool full of technical terms relevant to the problem at hand. But it also serves as a means of identifying members and folding them into the group's embrace while freezing out intruders. For instance, when computer programmer Ellen Ullman says a man's lovemaking makes her feel "like a user-exit subroutine," she is, among other things, making a profound declaration about her professional identity. Nontechnological bystanders may not have a clue what she is talking about, except for a pretty good hunch that this is an algorithm for lousy sex.

The Enron traders also had their own lingo. A "buck" was their term for $1 million and to "puke" meant to sell a position at a loss. Obscenities were a mark of membership in the club. When a timid young associate complained to Enron vice president Sherron Watkins that he had "never heard the term 'buttfuck' used in a business meeting before," Watkins icily suggested he go visit the trading floor, "meaning, 'I belong here—what about you?' "

"Everything was an inside joke," Cruver writes in his Enron memoir, *Anatomy of Greed,* "and the jokes would build and change and live for years." Cruver eventually figured out that you become part of a culture not by remarking on it but by conforming to it: "If you didn't get the joke, you laughed anyway, because that was the only way you could ever get 'invited' to the joke."

The willingness of almost everyone to conform to the group, and to go along with the joke, was of course ultimately the downfall of the company. It remains a lasting reminder of how powerful and sometimes dangerous the urge to surrender oneself to the group identity can be.

THE FAMILY WAY

The traditional way to create a sense of affinity in a group was through kinship. Nepotism comes naturally to primates. Like Murdochs and Agnellis and Toyodas in the human world, monkeys and apes routinely discriminate in favor of close kin. In many species, high-ranking mothers and grandmothers protect their youngsters when lowborn group members threaten them. They also interfere at playtime to ensure that young Lachlan Macaque and little Akia Vervet get their way. Strangely, even second-tier females give extra help to youngsters who are born to power. These proxy aunts apparently have a vested interest in preserving the stability of the traditional order.

The new generation thus grows up secure in the habit of defeating the family's subordinates, and dominance status gets passed on from one generation to the next. As a result, macaques, baboons, vervet monkeys, and other species all have family dynasties that, in the words of one biologist, "operate as corporate units." These dynasties appear to be based entirely on family power, not merit—not even merit in monkey terms of having bigger fangs or a surlier personality.

Nepotism (from the Latin word for "grandson" or "nephew") evolved as a natural behavior in part because it can be a shortcut to trust and cooperation in any group. We tend to be more comfortable around people like ourselves, and no one is more like us than our close kin. Hence the almost instantaneous affinity that often springs up when young cousins come together.

As kin selection theory points out, we also have a biological stake in the success of our close kin. We share half our genome with our siblings and children, a quarter with our nieces and nephews, and an eighth with our cousins. Helping them is a way to help ourselves, both in the evolutionary long term and (since they have an equal incentive to reciprocate) in the quarterly results here and now.

For instance, in one midwestern machine tool company, the founder's second wife sits on the board of directors, two of the founder's sons by his first wife are vice presidents of research and production, a nephew is company president, and a son-in-law runs the gear division. Nor is this nepotism confined to management. An unwritten company policy gives current employees' family members first shot as new hires. And most, if not all, of the women who work there are wives, or ex-wives, of employed husbands. "This is the way all small companies operate," says a family member.

Experts disagree about whether the family is a sound basis for building a business. Including kin can build trust and cooperation. But it can also demoralize nonkin who get unfairly excluded. Resentment is especially likely when a family uses supervoting stock or other devices to exert control and extract benefits beyond its actual ownership, as when Rupert Murdoch, who owns just 30 percent of News Corporation, put sons Lachlan, now thirty-three, in charge of the *New York Post* and James, thirty-one, in charge of British Sky Broadcasting.

One recent study in the *Journal of Finance* found that family-controlled public companies turn in significantly better market performance than nonfamily companies. The assumption is that the greater trust, cooperation, and personal investment in a family company make for better performance. But a more recent Harvard Business School working paper refines the previous analysis by separating family companies where the founder is still CEO from those companies where a descendant is CEO. Founder-led companies predictably outperform nonfamily companies in creating market value. But the Harvard analysis found that descendants "destroy value," especially in the second generation.

Apart from financial results, is there any scientific argument for having successive generations of the founding family wield power over a business? When the founder makes the decision to pass the company on to the kids, the implicit assumption is that they have also inherited his financial acumen. This may be so, but it's probably not genetic.

Even in animals, the case for nepotism appears to be based on nurture more than nature. Dogs, for instance, inherit the neural and olfactory equipment to be experts at smelling things. Dog owners have routinely used genetics, via selective breeding, to emphasize this trait in some breeds, including German shepherds. So it's not too surprising that when German shepherd puppies are reared with mothers trained in narcotics detection, 85 percent of them show an aptitude for narcotics detection work.

The surprising thing is that when the same puppies get raised by untrained mothers, only 19 percent of them show this aptitude. That is, even though the dogs may have a genetic predisposition for the family business, they pick up the family expertise mainly by hanging around with Mama and keeping their eyes open (or their nostrils). William S. Helton, a psychologist at Wilmington College in Ohio, suggests that this evidence about the importance of nurturing "might provide some insight into the phenomena of expertise running in human families, such as the musical skills of the Bachs or the mathematical skills of the Bernoullis."

The doggy comparison is not so far-fetched as it might sound to outraged Bach lovers. Like dog breeding, human mating among the founding families of great dynasties has often been a conscious effort at selective breeding. Parents have routinely maneuvered the next generation into marrying partners with desirable traits or families. In the early decades at the investment firm of Goldman Sachs, for instance, the founding families of Goldman and Sachs routinely intermarried. In the same collaborative spirit, the current CEO of Ford Motor is a great-grandson not just of Henry Ford, the auto company founder, but also of Harvey Firestone, the tire company founder.

As in dog breeding, some founding families have even attempted to concentrate their precious essence through generations of inbreeding among first and second cousins. The Rothschilds and the du Ponts are notable examples, both with extraordinary records of long-term financial success. Cousin marriages, and the close interweaving of business and family interests, are still common from North Africa into Central Asia.

It may not much matter what experts think about it. Nepotism is simply something people do, almost automatically. It is their nature. One recent study found that members of the founding family still play a role in a third of S&P 500 companies. Another study of almost three thousand corporations in East Asia found that two-thirds are controlled by families or individuals. And it can be hard to argue with the record of stability and longevity compiled by some family companies. For instance, the world's oldest hotel, the Hōshi Ryokan in Awazu, Japan, continues to be operated by the family that founded it in 717, almost thirteen hundred years ago.

WHAT AM I DOING HERE?

The powerful sense of belonging is most visceral in the small, tightly bonded group. But it's also possible to achieve a sense of collective identity at the company level. The social and emotional nature of the human animal is such that a company culture can motivate people even, paradoxically, in the sort of debased global economy where Junior's seat at the table, and Junior's replacement's seat, is always getting taken by the next Third World worker willing to accept an even lower wage.

Why would anyone ever feel loyalty to such an employer?

Just for the sake of argument, let's say that a company decides it can no longer afford to meet employees' basic survival needs. Maslow be damned. Sam Walton's ideas now rule, and the only way the company can meet the price demands of buyers at Wal-Mart is by offshoring its manufacturing to some destitute backwater. Even there it cannot pay enough for workers to feed or shelter their families.

Such a company should presumably also disregard the social needs of its employees, right? The relationship is a straight cash proposition, with no commitments on either side. It would be irrational to expect workers to feel loyalty, wouldn't it? But I have met auto industry employees living in cardboard shacks in the Ciudad Juarez dump, fifteen minutes from the American border, who nonetheless found a sense of identity in the workplace. Mostly it was identity with their fellow workers on the assembly line. But they also showed inklings of pride in the company.

Instead of ignoring or trying to suppress these social yearnings, any company will do better by asking how it can gratify them. One obvious way to do so is by giving workers a stake in the company culture. At the minimum, a brief introduction to the organization's goals and values helps give new employees a crucial sense of purpose and identity, beyond the simple mechanics of the job. These values should include a commitment to meet the worker's basic survival needs at some point in the foreseeable future, preferably before starvation sets in. But even in the meantime, the company culture gives workers something to which they can belong—not quite a family, but more like an echo of our old tribal cultures.

And it is entirely reasonable for even underpaid workers to want to be part of that. In the natural world, a meerkat, a wild dog, or a baboon leaving the group where he was born and attempting to join a new group often endures months of uncertainty and ill treatment. He hangs around on the fringes, hoping for a scrap of food when everybody else has eaten. He gets picked on and snarled at. But he takes the abuse and sticks with the group. Maybe he becomes more like the group, or maybe they just get used to him. Eventually the group opens its ranks and accepts him.

Even then, in many species, the individual may be denied sexual privileges and kept in a state of reproductive immaturity for years afterward. (A bit like young workers who can't afford to have children.) Putting up with all this is worth it, though, because the group offers safety against predators, occasional scraps of food, the consolation of social contact, the gratification of becoming part of something larger than oneself, and the chance that, someday, even the lowliest newcomer may rise to enjoy the full perquisites of power.

None of this justifies ill treatment of human workers. But it's worth at least being aware of how much company cultures function like cultures in the natural world. Ants, for instance, know the scent of their colony and recognize other ants accordingly. Chimpanzees live and die by the rules of the troop. Wolves run with the pack—and not just any pack, but their own pack. Whale pods and songbird groups have their distinctive dialects, which they learn growing up and which can shift over the course of their lives. It isn't so different from becoming part of the culture at LG, or Sanofi-Aventis, or Toyota.

SINGING THE COMPANY SONG

With social animals, it's almost never enough simply to be a member of the same species. The right song, smell, or dialect—or the right buzzword in a business context—identifies the individual as part of the same group, and this entrée can make an enormous difference to the individual's survival. Among sperm whales, for instance, clicking to the distinctive coda of the social unit seems to affect feeding success. Among cowbirds, young males sing randomly at first, but when they produce the local dialect, females respond with "copulation solicitation displays," a form of positive reinforcement that induces remarkably rapid learning in the males. And among mountain white-crowned sparrows, males singing the local dialect get more sex and more offspring.

What about us? Matsushita actually has a company song, and workers in Japan sing it daily. The effect on their sex lives is, alas, unknown. But it is standard human behavior to seek out the markings of cultural membership and, based on that, either to practice discreet avoidance or to relax into the warm bath of newfound kinship. ("We'll have a round of oxytocin, please.")

San Jose State University anthropologist Jan English-Lueck has spent more than a decade studying the ethnography of Silicon Valley savages. When two strangers from the technology industry meet at a local park with their children, she says, it's like "a Tuareg warrior and his entourage riding across the Western Sahara desert of North Africa" and meeting up with a stranger. Is he a fellow believer, perhaps even a distant kinsman? Or is he "someone who is liable to slit their gullets in the night[?] . . . The two strangers in Silicon Valley likewise trade seemingly idle talk that is actually in search of a common employer somewhere in their resumes, or a shared professional specialty."

The value of enlisting employees in the company culture ought to be apparent: When people share a set of communal beliefs and values, they tend to coordinate their work better, enjoy a greater sense of direction, and feel a deeper sense of loyalty and commitment. And yet plenty of organizations fail to help workers find their way into the culture. Maybe it just seems too obvious. Or maybe employers deem it a waste of time because it doesn't seem to contribute directly to the bottom line. Often they just don't understand their own culture. "Culture hides much more than it reveals," the anthropologist Edward T. Hall once declared, "and strangely enough what it hides, it hides most effectively from its own participants."

Whatever the reason, most employers seem to do a lousy job of communicating their culture. One administrator spent most of his career at a culture-conscious technology company where managers all got properly trained in the same methods and language, so they could communicate with one another and also understand what constituted success in their world. Then he switched to a job at one of the world's great universities, where the three-hundred-year-old culture is as rich as the ivy on the walls or the music pealing from the carillon tower every evening. But the administrator described his introduction to the university this way: "I was told where my office was. And I had to find it." It took him several years to learn the language and start to understand the culture, years spent far less productively than might otherwise have been possible.

CULTURAL BIRD-WATCHING

Ask around at any organization, and it's a safe bet that most employees won't have a clue about their company's fundamental purpose or why it's good for them to be a part of it. "I'm just the receptionist," they'll say with a shrug, or "I just work here." Better yet, walk around and ask the employees at your own company. Do it now. They will look at you as if you have lost your mind. But they may also admit that they have never gotten the crucial sense of belonging because no one bothered to tell them what it is they were supposed to belong to.

These are the sorts of things supposedly covered in company mission statements. But author Richard Saul Wurman once made an informal survey of thirty-six companies, phoning up and asking the person who answered the phone to identify the company mission statement. Only five could do so, and fewer than half could produce someone who could even lay hands on the mission statement. Among the responses Wurman got: "We have a very lengthy mission statement, but it's not for external publication," "I thought I had it on my computer, but I can't find it now," and "What's a mission statement?"

The Internet makes it easier to learn what a company thinks it is all about. But at many company Web sites, no mission statement is in evidence, or it takes multiple clicks to find it. And when you do find it, it is often insipid to the point of being meaningless: "With utmost respect to human values," goes one such fearless declaration, "we promise to serve our customers with integrity through innovative, value for money solutions, by *applying thought* day after day" (the italics are theirs).

The company in question, Wipro, a leading supplier of information technology outsourcing in India, apparently thinks that applying thought sets it apart from the rest of the corporate world. And maybe so. Still, it would be nice to know roughly the sorts of questions to which thought is being applied. (And "day after day"? Is it smart to enshrine the numbingly repetitious character of work life in your grand vision?)

In any case, written mission statements are less persuasive than what people experience face-to-face. At Enron, the mission statement publicly committed the company to "respect, integrity, communication and excellence," and executives fatuously handed out T-shirts with the RICE acronym. But by their conduct, the same executives soon let new employees know that their real mission was to steal gleefully from grandmothers.

Some companies may seem too big for employees to feel any emotional connection to the larger corporate culture. Wal-Mart must somehow communicate its grand purpose to 1.4 million workers around the world. And since almost half of them soon quit (the company's grand purpose perhaps not sufficiently including their own betterment), it must do so all over again with 600,000 new employees next year. At that scale, the company culture can easily slip out of control. True believers at Wal-Mart may think of themselves as price cutters to the world. They may live and breathe the religion of lower costs for consumers. But the public perception of the company has shifted, as a former CEO recently acknowledged, from endearing underdog to overbearing top dog.

Economic cycles and normal patterns of company growth can also ravage a company culture. JetBlue Airways still feels young and hip. But at American Airlines, once the proud elite, passengers have evidently become a nuisance. An unhappy flight attendant recently told the *New York Times* that the company's unofficial slogan is "If you can't afford the bus, fly us." She went on: "They don't complain about JetBlue or Southwest and other Wal-Marts of the sky because there they expect no frills. On the larger carriers, passengers are still under the impression that they will get playing cards, magazines, peanuts, meals, and pretty stewardesses at their beck and call. Ain't gonna happen."

What this culture seems to be saying (if we creep forward quietly, like a naturalist drawn to the dulcet countersinging of immaculate antbirds) is, "Siddown. Shaddup. Give us your money and don't expect us to be polite about taking it."

The bird-watching metaphor may be more useful than it sounds. Customers, business partners, and prospective employees all do better when they can scout out a company's culture before becoming enmeshed in it. But companies generally go out of their way to put on the best possible one-happy-family face. For instance, an executive assistant made a point during her job interview of asking about the culture at a California technology company where she was being recruited to work for the CEO. The senior vice president for human resources told her, "Oh, we don't have a company culture. We're all about getting the job done."

The executive assistant took the job and discovered too late that this should have been a warning signal. The company culture turned out to be "very tough, very backstabbing, and very hypocritical." She might have spared herself the next three years of unhappy employment by taking time to observe the culture beforehand. It's a simple matter of sitting in the blind and paying attention to how the animals behave—in this case, how people act in the company cafeteria, or how they look coming in to work. Are office doors open or closed? Are the hallways and common areas lively, or do people glance around furtively and hustle through? Do people chat across their cubicles or do they hunker down?

Richard Gallagher, a customer service consultant, once made the mistake of buying a laptop computer from a company "whose view of service was consistently shoddy at best." On a West Coast road trip, Gallagher discovered that the computer company shared a parking lot with one his client firms. "The next morning, I made it a point to come in bright and early, staked out a window, and just sat there and watched as employees from the two companies came in to work for the day. People from my client's firm walked tall, waved hello to people in the parking lot, and smiled and joked with each other as they walked in to work. Meanwhile, the other firm's employees trudged in silently, many with their eyes focused on the ground and their shoulders hunched. They looked like convicts marching off to jail."

WHEN GROUPS MEET

Company cultures naturally get more complex the longer you watch. Often there is no single corporate culture. One part of a company can be as lively as a duck pond in mating season, while another, just down the hall, feels like a mausoleum in winter.

Different cultures also sometimes clash, or complement each other, or blend together and shift to something entirely new. Understanding these cultural crosscurrents and figuring out how to work with them is everybody's basic survival challenge. If you know the culture and use the right words at the right time, you may just get that new stapler you covet. Use the wrong words and your $10 million budget request can end up in the oubliette, or you may get fired. There is nothing soft and fuzzy about it. Cultures translate directly into cash outcomes.

For example, the prestigious or powerful social connections of a company's managers may seem to have little to do with the business. They may never turn up in any financial report. But they tend to be public knowledge within a particular business community, and a 1993 study found that the badges of elite cultural status can provide protection in a takeover fight, making a high-status company more likely to prevail over a lower-status target, or more likely to extract a top price if the company becomes a target itself. A study of an Australian airplane merger found that employees of the lower-status partner also suffered more negative effects of the merger.

The cultural clash when such companies merge is often the amplified sound of separate prestige systems doing battle for survival. It's like two tribes coming together, each with its own chieftains and warriors. When General Motors acquired Ross Perot's Electronic Data Systems for $2.5 billion in the mid-1980s, for instance, GM CEO Roger Smith discovered that one of his new subordinates actually earned more than he did, an intolerable breach of hierarchical etiquette.

This kind of clash gets repeated all the way down the ladder and can take years to work itself out, in ways that are sometimes remorseless and more than a little demented. In the era before computers, for instance, two New Jersey newspapers merged. The news editor at the lesser paper was busted down to county editor in the combined operation, and he bitterly resented the rival who became the new managing editor. The two of them sat twenty feet apart in the open newsroom.

The county editor always knew well in advance what stories his reporters were planning to send in on the Teletype. His chief pleasure in life was to hang on to a big story till five minutes before deadline, when the managing editor had already painstakingly laid out page one. Then he would lift the Teletype copy, a ribbon of yellow paper roughly three feet in length, on the tip of his long scissors, and carry it outstretched to the managing editor's desk.

"Just got this from Paterson," he'd say, his head tilted at a forty-five-degree angle, the smoke from his Pall Mall drifting up into his left eye, the scissors held out like a sword. It was a perfect moment of ritualized aggression, which the entire newsroom watched with morbid fascination.

"Thought you might like it for page one," he'd say. Then he'd smile.

Unfortunately for the county editor, his reporters recognized that their future lay not with him but with the managing editor. In a bit of coalitionary maneuvering any chimp would readily recognize, they soon learned to phone the managing editor on the sly and give advance warning when they thought a story might rate front-page treatment. So the managing editor, with a two-inch hole ready on page one, always ignored the scissors pointed at his throat and remained scrupulously polite.

"Thanks very much, Al," he'd say.

THEM

If it's normal primate behavior to form a group identity and feel a powerful connection to it, the flip side is that we also tend to view anybody who is not "us" as "them." Knowing this, business leaders commonly focus on an enemy, or even invent one if need be, as a way to get employees to stop gnawing on one another and work in unison. When Lou Gerstner wanted to shock IBM out of its complacency, for instance, he did so by urging employees to get angry about the way three prominent rivals had "ripped away" market share: "This competitive focus has to be visceral, not cerebral. It's got to be in our guts, not our heads. They're coming into our house and taking our children's and our grandchildren's college money. That's what they're doing."

Worrying about the grandkids' college money may seem a little soft as a gut issue. But Gerstner was clearly trying to elicit the ancient feeling of threat and hostility between rival tribes.

Even within a company, it's common to create separate in-groups and encourage bias against the dominant culture. When businesses feel themselves getting big and stodgy, they often try to recapture their youthful zeal by creating separate entities called "skunkworks," where they put their iconoclasts to work. Apple did it to create the Macintosh, for instance, and members of the in-group were so elitist that they would not let the Apple II drones who were paying their salaries into the building.

A veteran of another such skunkworks, at Sun Microsystems, recalls, "Management told us we were not to talk to anyone about what we were doing, even other Sun employees. There were seminars where people would go and share their work. I could go, but I couldn't speak. It created a mystique. The people heading up the skunkworks were not entirely mature. They claimed they were doing it to keep corporate antibodies from attacking. But I suspect there may have been an element of 'I know something you don't know. . . .' " On the other hand, they did apparently know something, as the group produced the Java programming language, which proved hugely profitable for the company.

Us-and-them thinking is entirely natural. This isn't the same as saying it is entirely healthy, nor is it necessarily the smartest way for corporate leaders to go about creating a culture. Harvard primatologist Richard Wrangham characterizes what he calls in-group/out-group bias as the darker side of our inheritance from some murderous, chimplike, party-gang ancestor. It may often be "stupid and cruel," he writes, but it's also "perfectly expected in a species with a long history of intergroup aggression."

As part of his thesis about the origins of human violence, Wrangham suggests that we bond with our in-group essentially to become better at beating up out-groups. This is the sort of thing some of the more bloody-minded business leaders like to hear. It is, however, a highly arguable contention.

A closely bonded group can no doubt fight more ferociously. This may partly explain why in-groups and out-groups often face off far more forcefully than anyone intends. At one factory, for instance, the managers pitted two shifts against one another in a friendly competition. But the two shifts became so rivalrous that it led to unsafe work conditions.

At another plant in a midwestern farming community, a new manager unwittingly provoked in-group/out-group hostilities along racial lines. Many of the employees had worked there for decades, and they had built a solid reputation for the plant, which they had come to regard as their own. The tradition was for local technicians to run things, with the manager serving mainly to buffer them from headquarters. But the new manager was specifically assigned to introduce major changes, partly because constant innovation was a specialty of the corporation and partly to meet new globalization standards.

This might have been a recipe for conflict entirely by itself. But the community was largely white, and the new manager was a fast-track executive, African American, in his mid-thirties. When he started to meet resistance, he made the fatal mistake of appealing to his own most obvious in-group, asking the plant's handful of black workers to serve as his eyes and ears.

By the time a consultant came in to study the problem, the plant was out of control. The manager had dug in, black workers were taking his side, and white workers got red in the face talking about it. "He wouldn't have been in trouble, except for that spy network," said the consultant. "You can't do that, no matter what color you are. And once you bring in the racial element, it's like a stick of dynamite." The consultant made a confidential report to her contact at the corporate level. The company quickly yanked the manager and (apparently as a reward for his people skills) reassigned him to human resources, an effective dead end for a fast-track career.

NOT OUR KIND

It's a common tendency not just for in-groups and out-groups to fight but also for each to regard the other as not entirely human. In the language of psychologists, "encapsulation" describes the phenomenon of a group isolating itself from others and taking on a common mind-set. The members of the in-group share certain viewpoints, values, language, and other markers. People who don't dress or talk the "right" way can become almost invisible. It becomes easier to inflict pain on them without guilt. For instance, in the late 1980s, as GM was rapidly losing market share, CEO Roger Smith laid off thirty thousand assembly line workers without apparent regret.

Then a vice president suggested that GM was being slowed down by mediocre talent at the top and proposed firing 20 percent of upper management—roughly one hundred men earning at least $125,000 a year. "Smith wouldn't allow it," Doron P. Levin writes in *Irreconcilable Differences,* his account of the period. "He was convinced he could turn GM around without that sort of tumult." Why did Smith equate one hundred layoffs with tumult but not thirty thousand? Because the one hundred were "us," familiar faces in the company's fourteenth-floor inner sanctum. The thirty thousand assembly line workers were clearly "them."

US

All this suggests how hazardous the natural inclination to in-group/out-group thinking can be. There is, however, also ample reason to think that we form in-groups primarily for reasons other than beating up on out-groups. We do it because the need to belong is a universal trait of social animals like us. As a species, human beings are characterized by "obligatory interdependence," says social psychologist Marilynn B. Brewer, and group living represents "our fundamental survival strategy."

If we mean to prosper over the long term, "we must be willing to rely on others for information, aid, and shared resources," she writes, and we must be willing in turn to give information, aid, and resources. Both laboratory experiments and field studies indicate that the in-group is essentially a means of establishing this atmosphere of trust and mutual exchange.

Hating the out-group is almost irrelevant. For instance, Brewer made a study of thirty ethnic groups in East Africa, an area where stereotypes might encourage us to expect a high level of ethnic tension. She found that most test subjects ranked their own in-group more positively in terms of trustworthiness, obedience, friendliness, and honesty. But their view of out-groups wasn't correspondingly dim. In fact, a third of the groups rated at least one out-group more positively than the in-group in some categories. She found no evidence to support the traditional idea that out-groups always evoke negative attitudes. People tend, in other studies, to steer resources toward their own in-group. But when given the opportunity to harm out-groups more directly, they are reluctant to do so.

To put it another way, what makes employees love Procter & Gamble isn't the opportunity to beat up on Colgate. What makes people want to excel at Lands' End isn't their pathological loathing for L. L. Bean. And even if the new head of J. P. Morgan nurses an old grudge against his former boss at Citigroup, it's a safe bet his staffers won't get promoted by making voodoo dolls of the enemy. What matters are the relationships they form inside their own companies and the results they achieve as a consequence.

It may be fun now and then to get revved up against some archrival. But it can also quickly become ridiculous. In the pinkies-up world of tea drinking, for instance, Celestial Seasonings nursed a bitter grudge against rival R. C. Bigelow. At one point, salesmen at Celestial Seasonings placed little Bigelow tea mats in their urinals, to remind them of their target. Sales reps at a retreat received wooden mallets and handsome tins of Constant Comment, Bigelow's best-known product. Then, on cue, they used the mallets to hammer the tins flat.

Maybe it was motivational.

But sooner or later, at least one or two people must have looked around the room and thought, *Oh, my God, do I really work for this company?*

What makes people want to belong to a group is what goes on *inside* the group. *Do I know these people?* (Or perhaps, *Do I* want *to know these people?*) *Do they appreciate my work? Do they treat me fairly? Do I admire their work and can I learn from them? Do I have the opportunity to get ahead? Do they protect me? If I look out for them, will they also look out for me?*

These may seem like quintessentially human questions. But they come down to the same single factor that holds any group of social animals together. That is, do we trust one another enough?

One of the favorite images hanging in ranches and farms all across the Great Plains in the late nineteenth century was a painting by a Polish realist, Alfred von Kowalski-Wierusz, titled *Lone Wolf, February*. It depicted a solitary wolf standing on a snowy crest at night, its breath forming a blue plume in the bitter cold, looking down on a snug cabin warmed by a fire in the hearth. In the iconography of the frontier, the lone wolf came to stand for the heroic loner, who remains outside the comforts of civilization and refuses to be bound by its petty rules.

Unfortunately for this stalwart mythology, lone wolves are generally losers. People on the frontier ought to have known as much, as they were busily trapping, poisoning, and shooting wolves to the edge of extinction. But even in the unmolested natural world, the lone wolf is typically an outcast. Other members of his pack have driven him out, or he has wandered off, having failed to find a satisfactory place in the hierarchy.

It is still possible to regard the lone wolf as a hero, a pioneer heading off into unknown territory. But if he doesn't find a pack there, or form a pack of his own, he's likely to be a dead hero before too long. It's not just that lone wolves miss out on what one writer calls "the chatty comforts of wolf society," but they also cannot take on moose and other big game, which a pack working together handles routinely.

Humans are no less dependent on one another. Entrepreneurs, telecommuters, takeover artists, my-way-or-the-highway bosses—all of us, really—should pause every now and again, take a look at the people we work with, and then (moving past the moment of quiet despair) recollect this lesson: It's okay to be a lone wolf for a little while. It's good to get away from the bickering, the yapping, the flea-bitten fuzzy-mindedness, the nasty episodes of teeth-bared brawling. But the call of the wild is in truth a call back to our natural society. It's a call back to the troop, the tribe, the group. If we answer it, we are likely to discover that the greatest pleasure we can know in this life is to run with a good pack.

EPILOGUE
Leadership Lessons of Highly Effective Apes

We ride through life on the beast within us.

—Luigi Pirandello

Not long ago, a television production company where I have sometimes worked asked me to help pitch a documentary idea to a company in London's Leicester Square. We needed $750,000 to get the project off the ground, and we were only about halfway there. Commissioning producers in television are typically young, nervous, and unwilling to commit. They live in terror that a project will fail and bring down the withering scorn of the executive producer. But we needed to get a substantial share of the rest of our budget as a result of this meeting, or die.

The commissioning producer led us to a small conference room, decorated with various playful knickknacks attesting to the creative nature of the enterprise. One of them was a wind-up dog, which strutted down the table between us, barking and turning its head from side to side as if with interest. All of us smiled benignly and shared the requisite moment of child-like delight.

Then, by way of opening remarks, I said, "If you could just get it to piss on people's ideas, it could be an executive producer."

Everyone inhaled. Call it a gasp, if you must. And the project somehow never found its way into production.

So let me be the first to admit that applying animal analogies in the business environment can be hazardous, and it is entirely possible to take the whole thing just a little too far. Listeners often regard any comparison to animals as fundamentally insulting. Sometimes, *mea culpa,* it *is* insulting.

But so is looking in the mirror. And yet it sometimes pays to acknowledge the truth before our eyes. (I acknowledge, for instance, that I am still struggling with this whole business of being nice, even strategically.)

Let's resort to an animal analogy that happens to be highly popular among corporate types. At the *New York Times,* executives on the business side "sometimes award each other a small beanbag moose to recognize particularly probing questions." (The quote is from the *Wall Street Journal* and betrays a certain muted glee about the management-guru follies at its midtown rival. The *Journal* adds with labored sympathy that this sort of stuff makes *Times* staffers "from the news side cringe.")

The beanbag moose is a reference to one management consultant's sermonette about the "moose on the table." Or, in a current version geared to the booming Asian market, "the water buffalo on the table." And both versions probably derive from the more traditional notion of "an elephant in the room," used, for instance, by Alcoholics Anonymous as part of its twelve-step program. The idea is that people come to a dinner and realize that there is some damned big hairy beast or other standing on the table. But nobody talks about it. The topic is too big and embarrassing, and they'd rather pretend it's not there and hope it will just go away.

In the same manner, the sermonette suggests, companies often skirt big, ugly topics they are afraid to talk about out loud. (The sales manager's staff lives in terror because of her bad temper. Nobody wants to work with Tom in accounting because he smells bad. Thelma in procurement is never sober after 2 p.m. The company has no direction because the eighty-two-year-old founder won't acknowledge that he needs someone to take over in case he drops dead.) At many companies, the phrase "there's a moose on the table" has become code to stop the conversation and start talking about the gut issue.

But at all companies, if I may borrow the management consultant's device, human nature is the real moose on the table. Whatever else we may be, whatever divine spark we may possess, the inescapable reality is that human beings are also animals. We are apes, to be precise.

Science writer Matt Ridley puts it eloquently: "There is no bone in the chimpanzee body that I do not share. There is no known chemical in the chimpanzee brain that cannot be found in the human brain. There is no known part of the immune system, the digestive system, the vascular system, the lymph system or the nervous system that we have and chimpanzees do not, or vice versa. There is not even a brain lobe in the chimpanzee brain that we do not share."

Exploring this connection, getting to know both how we resemble other primates and how we differ, ought to give us insight into every aspect of our lives. And yet people profoundly resist the idea. Once, when I was interviewing a woman about mate choice in humans and other species, she stopped and said, "I'm sure you don't mean to suggest that people mate like animals." There was a long pause in the conversation. Perhaps for her, mating was a mingling of twinned souls, like columns of smoke braiding together, with New Age music playing in the background.

"Isn't the animal nature of the act a considerable part of its charm?" I inquired finally, and the conversation staggered to a close.

Acknowledging our animality, for instance in the context of sex, doesn't preclude affection, or intellect, or even spirituality. It is entirely possible to accept our animal nature and at the same time feel love. It is also possible, if you like, to accept our animal nature and still believe in God: God made primates, saw that they were good, and so made one primate lineage a little better than the rest. A lot better, if you will. (If we must be apes, let's at least be *splendid* apes.)

It is even possible to find an unexpected kind of spiritual comfort in accepting our animal nature, thanks largely to the more subtle view of animal social life that researchers have begun to build over the past few decades. We now know that, beneath the veneer of civilization, we humans are not necessarily killer apes. To say that our genes are "selfish" does not mean that we are not genetically designed to behave selfishly. Nor have we evolved to be as ruthless as Chicago gangsters.

We are simply social primates, endlessly working out the business of living together. It is difficult business, because, like chimpanzees, we are a quarrelsome species. But the sense of ourselves as members of a community may also come to us more naturally than we imagine because of our long evolutionary history of affiliation, cooperation, reconciliation, and morality.

So if we accept all that, what does it mean? Understanding our animal nature, how should we behave differently? Or to put it in management-guru terms, what are some key strategies of highly effective apes?

▪ *Do something to ease the animal skittishness of the people around you.* Every new assignment, every new face, every meeting is a visit to the watering hole, and the question in the back of everybody's mind is: *Am I going to get eaten alive here?* Smart bosses find ways to reassure subordinates that they will come out intact and perhaps even with their thirst quenched. But bosses often are not smart.

At Booz Allen, the consulting firm, a senior partner once gave a young associate just four days to prepare a major presentation for Citibank. "The associate got stuck with a huge amount of research," a colleague recalls. "It killed his weekend, he missed his kid's Little League championship, and he worked four days straight without sleep, crunching numbers.

"Monday morning, he and the senior partner go up to the boardroom at the Fifty-third Street building. The senior partner has his hand on the doorknob to go in, and the last words to the junior guy, in a gravelly voice, are, 'By the way, if you fuck this up, I'll cut your balls off.' Then he puts on a big smile and walks into the room."

And these are the consultants?

Even monkeys know that the best way to get someone to fight on your behalf is to groom him, to provide him with a sense of safety, comfort, and reciprocity. Even herd animals recognize that if they look out for one another, then maybe tomorrow they will not die.

▪ *Commit random acts of aggression at your own considerable risk.* And only against rivals with whom you expect never to work again. Being unpredictable may minimize the cost to you and maximize the cost to your victims. It forces them to remain constantly vigilant on all fronts against an attack that may occur anywhere, or not at all. This is what makes terrorism so effective. But when you destabilize a competitor in this fashion, with unpredictable attacks, you risk retaliation in kind. In any case, with customers, suppliers, and employees, a business should be utterly predictable, to encourage trust and cooperation.

- *Work with the crowd.* Bear in mind that we are *social* primates and become fully human only in groups. For most people, the modern workplace means having too little time to deal with too many groups. It is essential to give the people in any group three things: the chance to get to know one another, a clear sense of what they need to accomplish together, and the tools to do the job. Then kindly get out of the way. They can figure out the rest on their own.

Keep in mind that people in groups naturally imitate one another as part of the process of bonding. This is healthy, up to a point. The best thing a boss can do is give them something good to imitate, in the form of an open and yet ultimately decisive personality. (Almost as rare among corporate chieftains as among chimps, but it works.) A smart manager can also keep the group healthy by bringing in different points of view, to counter the innate tendency to groupthink. Bringing in people with different working styles may also serve as a subtle way to change the behavior of a group through emotional contagion.

- *Expect hierarchy.* When someone says, "We don't have any hierarchy around here," have the sense to smile deferentially and say, "Oh, that's so wonderful. Was that your idea?" But don't believe a word of it. Status competition and hierarchy are inescapable facts of primate life. Though we disparage them, they are also essential tools for encouraging high performance and for preserving domestic tranquility.

In every situation, know who is in charge and behave accordingly. If that person happens to be you, remember the "cookie monster experiment" and recognize how power automatically distorts your perspective. Also recognize that, in most situations, the unnecessary use of power is a mark of weakness.

One of the alpha chimps Frans de Waal studies is a bristling middleweight named Bjorn, after the tennis player. "He's a very hyped-up mean male who has come to the top I think by fighting dirty," says de Waal. Within the group, rival males generally fight by something like Marquis of Queensbury rules. But not Bjorn: "He makes injuries in the belly, in the scrotum, in the throat, places that are potentially dangerous."

It's possible Bjorn can survive at the top through sheer rottenness.

But he faces constant pressure from his closest rival, a larger, more sociable male named Socko, short for Socrates. Other subordinates show their reluctance to follow Bjorn by the pant-grunts with which subordinates acknowledge another chimp's superior status. "Bjorn has to work for his pant-grunts," says de Waal. "Socko gets his for free."

Which sort of boss would you rather be?

▪ *Share the fruits of success.* To get to the top, the conventional thinking goes, you need to be a self-serving brute. It's a law of nature, or so we tell ourselves. But behavioral research and anecdotal evidence indicate that alphas in the animal world also sometimes lead by being generous and considerate with their subordinates.

"Some years back, when my husband and I were visiting friends in India, a large male rhesus monkey appeared on the balcony outside the dining room," writes UCLA researcher Shelley E. Taylor. "Since rhesus males can be quite confrontational, we watched him from the safety of the kitchen. Once in the dining room, he looked around for food; spying a long loaf of sliced bread, he seized it, ran to the balcony, and climbed up into the trees and across the street to a field where his fellow monkeys were waiting. As we watched, they all sat down, our thief patiently unwrapped the bread, and one by one, he handed a slice to each member of the waiting group. As the alpha male, it was no doubt his role to undertake such risky activities, but in doing so, he helped to ensure his continued role as leader through his beneficence."

Frans de Waal has described similar behaviors among his chimpanzees: "Yet instead of dominants standing out because of what they take, they now affirm their position by what they give." This is, of course, the exact opposite of the behavior of supposedly civilized human CEOs who loot their companies and pocket the cash.

▪ *Trust, but check the numbers.* It's our nature to want to make friends and influence people. The personal connection fosters the sense of trust and reciprocity that is essential for doing business together. Even companies operating through the impersonal medium of the Internet thrive by somehow instilling loyalty in their customers.

So affiliate. But at the same time, beware of affiliation. It may help to recall that this contradictory dynamic has deep roots in nature: By night, chimpanzees huddle together in the treetops. By day, they engage in mutual grooming. But they also watch one another continually and protest unfair dealings with raucous barks of "Waaa!" It is much the same in any workplace. You need to develop trust, yet somehow stay alert for people who regard the personal connection primarily as a means for getting close enough to pick your pocket. You need to encourage coworkers to develop strong relationships with customers and suppliers while also staying alert to how these relationships may cost you money.

Cronyism is a natural hazard among social primates. Warren Buffett once nicely characterized the way CEOs put pals on the board of directors, especially on the committee that determines the CEO's own salary: "There is a tendency to put cocker spaniels on compensation committees, not Doberman pinschers." To help shareholders discover that the CEO, say, is cutting a sweetheart deal with his daughter's catering business, some countries require companies to reveal "related party transactions" or "interests of certain parties." For public companies in the United States, these filings are available at www.sec.gov. The Web site www.footnoted.org provides a useful guide to finding self-dealing and nest feathering in obscure SEC filings. It also regularly publishes its own juicy discoveries. Otherwise, gossip is our best means of self-defense.

- *Do the right thing.* Morality and fairness are not just nice ideas being taught in the more progressive business schools. They are part of our innate biological heritage. Chimps, like humans, live by a highly developed set of social rules. They display a keen appetite for punishing individuals who misbehave, empathy for victims of injustice, an interest in peacemaking after conflict, and, above all, an abiding concern with the maintenance of good relationships in the community. This is the rudimentary basis for morality, which may thus date back more than five million years, to the time when chimpanzees and humans shared a common ape ancestor.

Being good—being cooperative and conciliatory—remains the likeliest way to thrive in any social group, whether it's a chimpanzee troop or a brokerage house. Being bad may seem like a smart idea at the time. But evolution has made people hypervigilant about detecting cheaters, meaning somebody will eventually figure out what you are up to. Word will spread almost instantly via gossip (and the Internet). And if what you have done is a moral infraction or an act of treachery, negativity bias guarantees everyone will remember it for the rest of your life.

- *Pay attention to the nonverbal stuff.* Whether it's 50,000 or 250,000 years old, language is still basically a new product put on the market without adequate beta testing. Most of our emotional and neurological systems evolved long before we discovered the power of words, and most of our communication is still nonverbal. Moreover, compatibility problems sometimes occur between the old and the new systems. Our words say one thing. Our bodies say another. The body is almost always the more reliable indicator, not just for understanding other people, but also ourselves. Even when our prefrontal cortex is busily finding nice words to tell us one thing ("I think the boss and I are starting to get along better now"), our bodies say what we really feel ("He really hates my guts"). Learn to pay attention.

At the same time, beware, as body language can be relatively easy to manipulate. People have long since learned the importance of maintaining eye contact while telling lies, for instance. But facial expressions, and especially microexpressions, are less susceptible than large muscle movements to conscious control. Learn to use them to understand things people are not saying out loud.

- *Conflict can be healthy. Just don't let it get in the way of dinner.* The founder of a small Florida telecommunications company complained that his vice president of sales and his chief financial officer often let their mutual animosity interfere with the job: "I am constantly trying to keep them from killing each other." One was a short hockey-playing type from Philadelphia. The other was a tall, heavyset woman in her thirties, "tough with a capital *B*." Both were too productive to fire.

Physiologist Robert Sapolsky once described similar bad behavior in baboons chasing at top speed after a gazelle:

> *And they're gaining on it, and they're deadly. But something goes on in one of their minds . . . and he says to himself, "What am I doing here? I have no idea whatsoever, but I'm running as fast as possible,*

and this guy is running as fast as possible right behind me, and we had one hell of a fight about three months ago. . . . I'd better just stop and slash him in the face before he gets me." The baboon suddenly stops and turns around, and they go rolling over each other like Keystone Kops and the gazelle is long gone because the baboons just became disinhibited. They get crazed around each other at every juncture.*

But we humans are better than that, aren't we?

We may not much like the guy running just behind us. But we hold ourselves in line with the help of our superbly developed prefrontal cortex, the part of the brain, says Sapolsky, "that keeps us from belching loudly during the wedding ceremony, or telling somebody exactly what we think of the meal they made." It's the part of the brain that gets us to focus on long-term rewards. It's what enables us to set aside our differences in pursuit of common goals. It's what gets us to put up with delayed gratification, sometimes for years.

But not always.

The Florida CEO tried reasoning with his quarrelsome subordinates. Then he resorted to embarrassment, writing a letter to an advice column asking how to deal with the problem. He dropped the published letter on the guilty parties' desks with instructions to come up with a solution. For a little while, they became more civil. But then one day a smirk, a rolling of the eyes, or some disparaging remark set off a cell assembly in the brain. It hardly matters who said what. The old animal emotions came surging up, and instantly, they were slashing each other in the face again and rolling around like Keystone Kops.

Finally, the CEO said, "Stop the fighting, or I'll fire you both."

This dire warning sank into their prefrontal cortices.

No job. No paycheck. Home alone. Peddling resumés. And they began to work together peacefully again. (At least for now.)

This is what it means to be human. We marshal the vast powers of the rational mind and struggle to ride gracefully on the tide of underlying animal emotions. Our unruly impulses surge continually to the surface. We bicker over territory, engage in status rivalries, lust after colleagues, exert dominance over them, suffer slights, nurse grudges, and give in to fear.

Then we pull ourselves together again and get back in the race. We are not just a bunch of baboons, after all.

We are human, and we know how to stay focused on the gazelle.

For updates and additional information, visit www.apeinthecorneroffice.com.

1 YES, IT IS A GODDAMN JUNGLE OUT THERE

p. 1 Animals in the wild: Martel, Y. (2001), *Life of Pi,* New York: Harcourt.

p. 2 Chimps and us: de Waal, F. (2001), *The Ape and the Sushi Master: Cultural Reflections by a Primatologist,* New York: Basic Books, p. 71.

p. 3 "If you aren't the lead dog . . .": Bryce, R. (2002), *Pipe Dreams: Greed, Ego, and the Death of Enron,* New York: Public Affairs, p. 118.

p. 4 Elephant tap dance: *Business Week,* October 8, 1986.

p. 4 Weasels: Adams, S. (2003), *Dilbert and the Way of the Weasel,* New York: HarperBusiness.

p. 4 Fighting or grooming: de Waal, F. (1997), "The chimpanzee's service economy: Food for grooming," *Evol. & Human Behav.* 18:375–86.

p. 6 Staying small at W. L. Gore: *Sales and Marketing Management,* April 2003, p. 32.

p. 7 Dogs and soldiers: Bernhard, J. G., and Glantz, K. (1992), *Staying Human in the Organization: Our Biological Heritage and the Workplace,* Westport, CT: Greenwood, p. 47.

p. 7 "I don't do feelings": *Business Week,* July 26, 2004.

p. 7 MBA financial analysts at $800 a month: *Harvard Business Review,* February 2004.

p. 8 The code: Ullman, E. (1997), *Close to the Machine,* San Francisco: City Lights Books, p. 4.

p. 9 Sniffing the air and snorting: ibid., pp. 39–40.

p. 9 Rats specifically bred to display a high level of anxiety: Francis, D. D., Meaney, M. J. (1999), "Maternal care and the development of stress responses," *Curr. Opin. Neurobiol.* 9, pp. 128–34.

p. 10 Evolution of a smile: Schmidt, K. L., and Cohn, J. F. (2001), "Human facial expressions as adaptations: Evolutionary questions in facial expression research," *Yearbook of Physical Anthropology* 44, pp. 3–24.

p. 11 Multiple human smile types: Ekman, P. (2001), *Telling Lies,* New York: Norton, pp. 149–61.

p. 11 Matsui's New York debut: *Hartford Courant,* April 9, 2003.
p. 11 Women are better at smiling: Schmidt and Cohn (2001).
p. 11 Men on average 15 percent larger: de Waal (2001), p. 35.
p. 12 Thicker smile muscles in women: Schmidt and Cohn (2001), p. 8.
p. 12 Our forgotten history: Waldron, D. A. (1998), "Status in Organizations: Where Evolutionary Theory Ranks," *Managerial and Decision Economics,* 19 (7/8), pp. 505–20.
p. 12 Blurring the line between zoo and workplace: de Waal (1997); de Waal, F., and Berger, M. L. (2000), "Payment for labour in monkeys," *Nature* 404, p. 563.
p. 13 The Arioi in Tahiti: Stephenson, K. (1992), "How to Lead People," *Vital Speeches* LIX, 5 (December), pp. 138–41.
p. 13 The appetite for a payday: Knutson, B., et al (2001), "Anticipation of increasing monetary reward selectively recruits nucleus accumbens," *J. of Neuroscience* 21:RC159.
p. 13 How the brain responds to unpredictable rewards: *Natural History,* September 2003.
p. 13 Cortisol research: Sapolsky, R. (1994), *Why Zebras Don't Get Ulcers,* New York: Freeman.

2 NICE MONKEY

p. 15 Survival of the fittest: Spencer, H. (1873), *Principles of Ethics,* 2 vols., Indianapolis: Liberty Classics, 1978.
p. 15 Darwin letter: Rachels, J. (1990), *Created from Animals: The Moral Implications of Darwinism,* New York: Oxford University Press, p. 62.
p. 17 John Kay's 1998 speech: www.johnkay.com/print/133.html.
p. 17 The social responsibility of business: Friedman, M. (1970), *New York Times Magazine,* September 13.
p. 17 Mother Teresa: *Forbes,* February 2, 2004.
p. 19 *Guanxi:* English-Lueck, J. A. (2002), *Cultures@SiliconValley,* Stanford: Stanford University Press, p. 178.
p. 20 The good associate: Rolfe, J., and Troob, P. (2000), *Monkey Business,* New York: Warner, pp. 146–47.
p. 21 Family firms: *Family Business Review,* summer 1996; a 2004 list of top family companies worldwide is at http://www.familybusinessmagazine.com/topglobal.html.
p. 22 The affiliation instinct: The tentative nature of the discussion is suggested by the title of one recent paper: Silk, J. B. (2002), "Using the 'F' word in primatology." *Behavior* 139 (2–3), pp. 421–46, "F," that is, for "friendship." And "the 'L' word" turns up in Sapolsky (1994), p. 97.
p. 22 Social support and survival: Seeman, T. E. (1996), "Social ties and health: the benefits of social integration," *Annals of Epidemiology* 6 (5), pp. 442–51.

p. 23 People around us influence our biochemistry: DeVries, A. C., et al (2003), "Social modulation of stress responses," *Physiology and Behavior* 79 (3), pp. 399–407.

p. 23 Tend and befriend: Taylor, S. E. (2000), *The Tending Instinct,* New York: Times Books.

p. 24 Szalai, A. (1972). *The Use of Time: Daily Activities of Urban and Suburban Populations in Twelve Countries,* The Hague: Mouton.

p. 24 An experiment at Emory: Rilling, J. K., et al. (2002), "A neural basis for social cooperation," *Neuron* 35, pp. 395–405.

p. 24 Natalie Angier: *New York Times,* July 23, 2002.

p. 25 Prairie voles: Insel, T. R., and Young, L. J. (2001), "The neurobiology of attachment," *Nature Reviews Neuroscience* 2, pp. 129–36; Young, L. J., and Wang, Z. (2004), "The neurobiology of the pair bond," *Nature Neuroscience* 7, pp. 1048–54.

p. 27 Neurophysiology of trust: McCabe, K., et al. (2001), "A functional imaging study of cooperation in two-person reciprocal exchange," *Proc. of the Nat. Acad. of Sciences* 98 (20).

p. 29 Camp David: Carter, J., *Keeping Faith,* Fayetteville: University of Arkansas Press, p. 399.

3 BEING NEGATIVE

p. 32 T. H. Huxley included his description of mankind in a letter dated February 10, 1895, and added to these negative traits that human nature also included "an angel bobbing about unexpectedly like the apple in the posset."

p. 34 Negativity and the paleness of comforts: Rozin, P., and Royzman, E. B. (2001), "Negativity bias, negativity dominance, and contagion," *Personality and Social Psychology Review 5,* pp. 296–320.

p. 34 Speedy differentiation between positive and negative: Smith, N. K., et al. (2003), "May I have your attention please: Electrocortical responses to positive and negative stimuli," *Neuropsychologia* 41, pp. 171–83.

p. 35 Separate negativity and positivity systems, and their evolution on the savanna: Cacioppo, J. T., et al. (2004), "The affect system: What lurks below the surface of feelings?" in Manstead, A.S.R., et al. (eds.), *Feelings and emotions: The Amsterdam conference,* pp. 221–40, New York: Cambridge University Press.

p. 36 Organ donors: The Gallup Organization, Inc. (1993), Survey for the Partnership for Organ Donation and Harvard School of Public Health, March 25–26, available at http://www.transweb.org/reference/articles/gallup_survey/gallup_index.html.

p. 37 Irrational economic decisions: Kahneman, D., et al. (1990), "Experimental tests of the endowment effect and the Coase theorem," *J. of Political Economy* 98 (6), pp. 1325–48.

p. 37 No negativity bias for intelligence: Skowronski, J. J., and Carlston, D. E.

(1992), "Caught in the act: When behaviors based on highly diagnostic behaviors are resistant to contradiction," *European J. of Soc. Psychology* 22, pp. 435–52.

p. 38 Oracle Dumpster-diving: *Wall Street Journal,* June 29, 2000.

p. 38 Procter & Gamble versus Unilever: *Fortune,* August, 2001; *Wall Street Journal,* September 7, 2001.

p. 39 Larry Ellison quote: Symonds, M. (2003), *Softwar,* New York: Simon and Schuster.

p. 39 Citibank: Brandenburger, A., and Nalebuff, B. (1996), *Co-opetition,* New York: Doubleday.

p. 41 Chihuahua quote: Enrico, R., and Kornbluth, J. (1986), *The Other Guy Blinked,* New York: Bantam, p. 105.

p. 41 EDS and Losada: Losada, M., and Heaphy, E. (2004), "The role of positivity and connectivity in the performance of business teams: A nonlinear dynamics model," *American Behavioral Scientist* 47 (6), pp. 740–65.

p. 43 Conflict in the early stages is good: Jehn, K. A., and Mannix, E. A. (2001), "The dynamic nature of conflict: A longitudinal study of intragroup conflict and group performance," *Acad. of Management J.* 44 (2), pp. 238–51.

p. 43 Acknowledgement and productivity: Luthans, F., and Youssef, C. M. (2004), "Human, social, and now positive psychological capital management: Investing in people for competitive advantage," *Organizational Dynamics* 33 (2), pp. 143–60.

p. 43 Marriage and positivity: Gottman, J., and Levenson, R. W. (2000), "The timing of divorce: Predicting when a couple will divorce over a 14-year period," *J. of Marriage and the Family* 62, pp. 737–45.

p. 47 Davidson, R. J., et al. (2000), "Emotion, plasticity, context and regulation: Perspectives from affective neuroscience," *Psychological Bulletin,* 126, pp. 890–906.

p. 49 Promega experiment: Davidson, R. J., et al. (2003), "Alterations in brain and immune function produced by mindfulness meditation," *Psychosomatic Medicine 65,* pp. 564–70.

p. 50 Medtronics and Bill George: *The Economist,* September 2, 2002; *Business Week,* August 22, 2003.

p. 53 Adelie penguins: http://scilib.ucsd.edu/sio/nsf/journals/peter97.html.

4 ROUGH BEASTS

p. 55 Craig Barrett quote: *Business Week,* June 15, 1997.

p. 55 Fear: Grove, A. (1996), *Only the Paranoid Survive,* New York: Doubleday, pp. 117–19.

p. 56 The Intel brawl: Jackson, T. (1997), *Inside Intel,* New York: Dutton.

p. 56 Bat-wielding executive: *Business Week,* June 15, 1997.

p. 59 Manager see, manager do: *Business Week,* April 3, 1995.

p. 59 Yeroen at Boeing: The *Hartford Courant,* March 13, 2005.

p. 60 Machiavelli at the zoo: de Waal, F. (1982), *Chimpanzee Politics,* New York: Harper and Row, p. 19.

p. 60 Maneuvering: ibid., p. 113.

p. 60 Nikkie and Yeroen: ibid., pp. 47–48.

p. 61 Luit's game face: ibid., p. 133.

p. 61 *Chimpanzee Politics* on GOP reading list: *Business Week,* April 3, 1995.

p. 61 Memo about "bizarre, twisted Democrats": Molly Ivins in *Hartford Courant,* August 8, 2003.

p. 62 Gingrich as Yeroen to Rumsfeld: *Wall Street Journal,* April 23, 2003.

p. 62 Human violence: Wrangham, R., and Peterson, D. (1996), *Demonic Males,* New York: Houghton Mifflin.

p. 63 Another 'Sure, boss' meeting: *Harvard Business Review,* January 2003.

p. 65 Conflict is normal: de Waal, F. (2000), "Primates: A natural heritage of conflict resolution," *Science* 289 (5479), pp. 586–90.

p. 66 Weaning as the first negotiation, and the conflict resolution model: Aureli, F., and de Waal, F. (eds.) (2000), *Natural Conflict Resolution,* Berkeley: University of California Press, pp. 26–28.

p. 67 Intrauterine sibling cannibalism: Gilmore, R. G. et al. (2005), "Oophagy, Intrauterine Cannibalism and Reproductive Strategy in Lamnoid Sharks," in Hamlett, W. C. (ed.), *Reproductive Biology and Physiology of Chondrichthyes Sharks, Batoids, and Chimaeras.* Enfield, NH: Science Publishers, Inc.

5 DONUT DOMINANCE

p. 68 Churchill on pigs: Humes, J. C. (1995), *The Wit and Wisdom of Winston Churchill,* New York: Perennial, p. 6.

p. 68 Get the machine guns ready: Langley, M. (2003), *Tearing Down the Walls,* New York: Free Press, p. 297.

p. 70 Jamie Dimon at Bank One: *Fortune,* July 22, 2002.

p. 70 Whose blood does Jamie want: *New York Times,* January 15, 2004.

p. 71 Why does somebody always have to win: *Wired,* August, 1995.

p. 71 Samuel Johnson: Boswell, J. (1776), *Life of Samuel Johnson;* also updated for political correctness in Wrangham, R., and Peterson, D. (1996), *Demonic Males,* New York: Houghton Mifflin, p. 191.

p. 71 Dominance at Stanford: Barchas, P. (1984), *Social Hierarchies,* Westport, CT: Greenwood Press, pp. 25ff.

p. 71 Social hierarchies among children: Barkow, J. H. (1975), "Prestige and Culture: A Biosocial Interpretation," *Current Anthropology* 16, 4, pp. 553–55.

p. 71 Tea ladies as catering supervisors: *Guardian* (UK), April 18, 2000.

p. 72 Deutsche Bank in heaven: *Independent* (UK), February 8, 1994.

p. 72 Hollywood hierarchies: *New York Times,* April 28, 2002.

p. 73 CEOs over six feet tall: Etcoff, N. (2000), *Survival of the Prettiest,* New York: Doubleday, p. 173.

p. 74 Male obsession with rank: Wrangham and Peterson (1996), p. 191.

p. 75 Maslow coins "dominance drive" and "self-esteem": Cullen, D. (1997), "Maslow, monkeys, and motivation theory," *Organization* 4 (3), pp. 355–73; de Waal, F. (1996), *Good Natured,* Cambridge, MA: Harvard University Press, p. 99.

p. 75 Primate work as Maslow's foundation: Hoffman, E. (1988), *The Right to be Human,* Los Angeles: Tarcher, p. 49.

p. 76 Chimp group behavior in wild: de Waal, F. (2001), *The Ape and the Sushi Master: Cultural Reflections by a Primatologist,* pp. 188–90.

p. 77 The alpha takes the lead: Cheney, D. L., and Seyfarth, R. M. (1990), *How Monkeys See the World,* Chicago: University of Chicago Press, pp. 47–48.

p. 78 IBM Kremlinology: Gerstner, L. (2002), *Who Says Elephants Can't Dance?* New York: HarperBusiness, p. 191.

p. 78 Increased fighting when rank uncertain: de Waal, F. (1982), *Chimpanzee Politics,* New York: Harper and Row, p. 118.

p. 78 A stable pecking order lays more eggs: Guhl, A. M., and Allee, W. C. (1944), "Some measurable effects of social organization in flocks of hens," *Physiol. Zool.* 27, pp. 320–47.

p. 79 Slotow, R., et al. (2000), "Older bull elephants control young males," *Nature* 408, pp. 425–26.

p. 79 Respect helps civilize alpha: Emerson, R. M. (1962), "Power dependence relations," *American Sociological Review* 27, pp. 31–41.

p. 79 Bad alphas up a tree: de Waal, F. (1982), p. 56.

p. 79 Lou Pai and the $45,000 weekend flights: Bryce, R. (2002), *Pipe Dreams: Greed, Ego, and the Death of Enron,* New York: Public Affairs, pp. 188, 209, 264.

p. 80 Kinder economies: *Wall Street Journal,* April 12, 2004.

p. 80 Pilots cleaning planes: *Fast Company,* May 2004, p. 75.

p. 80 Wanting a better parking space: Lutz, R. A., (2003), *Guts,* New York: Wiley, p. 183.

p. 81 Counting ceiling tiles: Welch, J., *Jack: Straight from the Gut* (2001), New York: Warner, p. 49.

p. 81 Hating hierarchy: *Business Week,* May 28, 1998.

p. 81 Even Chainsaw Al hates hierarchy: Dunlap, A. (1996), *Mean Business,* New York: Fireside, pp. 76–77.

p. 82 The U-turn in social evolution: Knauft, B. (1991), "Violence and sociality in human evolution," *Current Anthropology* 32, pp. 391–428.

p. 83 Ntologi: Nishida, T., et al (1992), "Meat sharing as a coalition strategy by an alpha male chimpanzee?" in Nishida, T., et al. (eds) *Topics in Primatology,* v. 1, *Human Origins,* pp. 159–74. Tokyo: Tokyo University Press.

p. 83 Phil Knight's philanthropy: *New York Times,* April 25, 2000; Associated Press, September 26, 2001.

p. 83 Hopi hierarchies: Boone, J. L., and Kessler, K. (1999), "More status or more children: Social status, fertility reduction, and long-term fitness," *Evolution and Human Behavior* 20, pp. 257–77.

p. 84 No egalitarian societies: Flanagan, J. G. (1989), "Hierarchy in simple 'egalitarian' societies," *Annual Review of Anthropology* 18, pp. 245–66.
p. 84 Andygrams: Jackson, T. (1997), *Inside Intel,* New York: Dutton, pp. 221–22, with amendments directly from the executive.
p. 84 Lassergrams: *Business Week,* April 19, 2004.
p. 85 John Chambers' office: *Wall Street Journal,* January 14, 2004.
p. 85 A critical view of Chambers: Young, J. S. (2001), *Cisco Unauthorized,* pp. 196–97.
p. 85 Where does power lie: Leavitt, H. J. (2003), "Why hierarchies thrive," *Harvard Business Review,* March, pp. 97–102.
p. 88 Clownfish: Buston, P. M. (2003), "Size and growth modification in clownfish," *Nature* 424, pp. 145–46.

6 TOOTH AND CLAW

p. 89 Territoriality at the BBC: Wyatt, W. (2003), *The Fun Factory: A Life in the BBC,* London: Aurum Press, p. 7.
p. 90 Sandy vs. Lou: Langley, M. (2003), *Tearing Down the Walls,* New York: Free Press, p. 73.
p. 91 Visitors to the Oval Office: *New York Times,* August 9, 2004.
p. 91 Nikkie-Luit reconciliation: de Waal, F. (2003), *My Family Album,* Berkeley: University of California Press, pp. 84–85; see also de Waal, F. (1982), *Chimpanzee Politics,* New York: Harper and Row.
p. 93 Al Capone: Kobler, J. (1993), *Capone: The Life and Times of Al Capone,* New York: Da Capo Press, p. 17.
p. 93 Skilling's bad language: Bryce, R. (2002), *Pipe Dreams: Greed, Ego, and the Death of Enron,* New York: Public Affairs, pp. 268–69; for a symptomatically different inside reaction, see Cruver, B. (2002), *Anatomy of Greed: The Unshredded Truth from an Enron Insider,* New York: Avalon, pp. 54–55.
p. 94 Reaction to an angry face: Sackett, G. P. (1966), "Monkeys reared in isolation with pictures as visual input: Evidence for an innate releasing mechanism," *Science* 154, pp. 1470–3.
p. 94 Durango as savage jungle cat: Bradsher, K. (2002), *High and Mighty: SUVs—The World's Most Dangerous Vehicles and How They Got That Way,* New York: Public Affairs, p. 99.
p. 95 Death Star: Cruver (2002), p. 10.
p. 95 Fastow's predators: Bryce (2002), p. 225.
p. 95 Playing darts at K-Mart: *Detroit Free Press,* July 2, 2002.
p. 96 Dueling desk plaques: *Times* (UK), March 29, 2003.
p. 98 Overcoming defensive bias: Sherman, D. K., and Cohen, G. L. (2002), "Accepting Threatening Information: Self-Affirmation and the Reduction of Defensive Biases." *Curr. Dir. in Psych. Sci.,* 11 (4), August.
p. 99 Stress and surrender: Mazur, A. (1985), "A biosocial model of status in face-to-face primate groups," *Social Forces,* December, pp. 377–402.
p. 101 Speed of amygdale response: Smith, N. K., et al. (2003), "May I have your

attention please: Electrocortical responses to positive and negative stimuli," *Neuropsychologia* 41, pp. 171–83.

p. 102 Testosterone overview: Mazur, A., and Booth, A. (1998), "Testosterone and dominance in men," *Behavioral and Brain Sciences* 21, pp. 353–63. Testosterone and career choice: Dabbs, J. M. Jr., et al. (1998), "Trial lawyers: Blue collar talent in a white collar world," *J. of Applied Social Psychology* 28, pp. 84–94.

p. 103 Testosterone in World Cup fans: Bernhardt, P. C., et al. (1998), "Changes in testosterone levels during vicarious experiences of winning and losing among fans at sporting events," *Physiology and Behavior* 65, pp. 59–62.

p. 103 Trained losers: Aureli, F., and de Waal, F. (eds.) (2000), *Natural Conflict Resolution,* Berkeley, CA: University of California Press, p. 82.

p. 104 Boot camp cortisol: Hellhammer, D. H., et al. (1997), "Social hierarchy and adrenocortical stress reactivity in men," *Psychoneuroendocrinology* 22 (8), pp. 643–50.

p. 104 Atrophy in the hippocampus: Davidson, R. J. (2000), "Affective style, psychopathology, and resilience: Brain mechanisms and plasticity," *American Psychologist* 55, pp. 1196–214.

p. 104 Lock 'em up: Bryce (2002), pp. 230–33.

p. 105 Verbal abuse at Gap factories: *Wall Street Journal,* May 12, 2004.

p. 106 Submission after being outstressed: Mazur (1985), p. 388.

p. 106 Always concur: Rolfe, J., and Troob, P. (2000), *Monkey Business,* New York: Warner, pp. 227–28.

p. 107 Approach and inhibition: Keltner, D., et al. (2003), "Power, approach, and inhibition," *Psychological Review* 110 (2), pp. 265–84.

p. 107 Power as the fundamental concept in social science: Russell, B. (1938), *Power: A new social analysis,* London: Allen and Unwin.

p. 109 Urinating in the corner sink: *Wall Street Journal,* July 19, 2004.

p. 112 Coalitions and Supreme Court decisions: Keltner et al. (2003).

p. 112 AT&T's costly cable spree: *Wall Street Journal,* May 26, 2004.

p. 113 Ask the cage cleaner: Sapolsky, R. (1994), *Why Zebras Don't Get Ulcers,* New York: W. H. Freeman, p. 281.

p. 113 Gender, power, and social awareness: Henley, N. M., and LaFrance, M. (1984), "Gender as culture: Difference and dominance in nonverbal behavior," in Wolfgang, A. (ed.), *Nonverbal behavior: Perspectives, applications, intercultural insights,* Lewiston, NY: C. J. Hogrefe.

p. 114 A BBC chairman's dalliances: Wyatt (2003), p. 142.

7 BENDING THE KNEE

p. 116 The black dog: Chalmers, R. (2003), *Who's Who in Hell,* New York: Grove.

p. 116 Hollinger's report to the SEC, dated August 30, 2004, is available at http://www.sec.gov/Archives/edgar/data/868512/000095012304010413/y01437exv99w2.htm.

p. 117 Difficulty getting back Roosevelt papers: *New York Times,* February 16, 2004.
p. 119 The shiver of subservience: *Vanity Fair,* May 12, 2004.
p. 120 Pant-grunting: de Waal, F. (1982), *Chimpanzee Politics,* New York: Harper and Row, p. 87.
p. 120 Appeasement behaviors: Keltner, D., et al. (1997), "Appeasement in human emotion, personality, and social practice," *Aggressive Behavior* 23, pp. 359–74; see also Anderson, C., and Berdahl, J. L. (2002), "The experience of power: Examining the effects of power on approach and inhibition tendencies," *J. of Personality and Social Psychology* 83, pp. 1362–77.
p. 120 Meetings reinforce hierarchy: *Wall Street Journal,* May 19, 2004.
p. 122 Submissive language: Lakoff, R. (1975), *Language and Woman's Place,* New York: Harper and Row.
p. 122 Feet like strawberry shortcake: *Training Magazine,* July 1999.
p. 122 The workplace as a Skinner box: *Talk of the Nation,* National Public Radio, June 19, 2002.
p. 123 Dissected testicles: Maccoby, M. (1976), *The Gamesman,* New York: Simon and Schuster, p. 82.
p. 124 Big boss X: Lutz, R. A. (2003), *Guts,* New York: Wiley.
p. 124 The power signal: Gregory, S. W., and Webster, S. W. (1996), "A nonverbal signal in voices of interview partners effectively predicts communication accommodation and social status perceptions," *J. of Personality and Social Psychology* 70, pp. 1231–40.
p. 127 Serotonin and power: Raleigh, M. J., et al. (1984), "Social and environmental influence on blood serotonin in monkeys," *Archives of General Psychiatry* 41, pp. 405–10.
p. 128 Generalizing to humans: Masters, R. D., and McGuire, M. T. (eds.) (1994), *The Neurotransmitter Revolution: Serotonin, Social Behavior, and the Law,* Carbondale: Southern Illinois University Press, p. 130.
p. 128 The importance of greetings: de Waal (1982), p. 118.
p. 129 Don Tommy: *New York Magazine,* March 3, 2003.
p. 132 A chorus of agreement to former dumb idea: Lutz (2003), p. 181.
p. 133 Frodo's lip chewing: *National Geographic,* December 1995.
p. 133 Watching Ross Perot's left ear: Levin, D. (1989), *Irreconcilable Differences,* Boston: Little, Brown, p. 103.
p. 133 The boss's cigar: Langley, M. (2003), *Tearing Down the Walls,* New York: Free Press, p. 44.
p. 133 Being the right-hand man: Dunbar, R. (1996), *Grooming, Gossip, and the Evolution of Language,* London: Faber, pp. 136–37.
p. 134 Offering first-class seat to the boss: *New York Times,* March 11, 2003.
p. 136 Marcos, and upward grooming by subordinates: Cheney, D. L., and Seyfarth, R. M. (1990), *How Monkeys See the World,* Chicago: University of Chicago Press, pp. 42, 71.
p. 137 James Truman grooms Newhouse: *New York Times,* October 26, 2003.
p. 139 Postural echo: Morris, D. (1979), *Manwatching,* New York: Abrams, pp. 83–85.

p. 140 Appeal aggression: de Waal, F. (2003), "Darwin's legacy and the study of primate visual communication," in Ekman, P., et al. (eds.), *Emotions Inside Out: 130 Years After Darwin's* The Expression of the Emotions in Man and Animals, New York: New York Academy of Sciences.

8 CHATTER IN THE MONKEY HOUSE

p. 141 Gossip poisons business: *Workforce,* July 2001.

p. 141 Gossip for unity, morals: Emler, N. (2001), "Gossiping," in Robinson, W. P., and Giles, H. (eds.), *The New Handbook of Language and Social Psychology,* New York: Wiley, p. 319.

p. 143 Ban water-cooler conversations: *New York Times,* December 28, 2003.

p. 143 Grapes, cucumbers, and fairness: Brosnan, S. F., and de Waal, F. (2003), "Monkeys reject unequal pay," *Nature* 425, pp. 297–99.

p. 144 Gossip in the dining hall: Dunbar, R. (1996), *Grooming, Gossip, and the Evolution of Language,* London: Faber, p. 123.

p. 145 Language evolved for gossip: ibid., p. 79.

p. 145 Energy for the brain: ibid., p. 124; and Dunbar, R. (1992), "Neocortex size as a constraint on group size in primates," *J. of Human Evolution* 20, pp. 469–93.

p. 146 Using positive words: Boucher, J., and Osgood, C. E. (1969), "The Pollyanna hypothesis," *J. of Verbal Learning and Verbal Behavior* 8, pp. 1–8.

p. 146 Positive first in word pairs: Rozin, P., and Royzman, E. B. (2001), "Negativity bias, negativity dominance, and contagion," *Personality and Social Psychology Review* 5, pp. 296–320.

p. 146 Only 5 percent of gossip is negative: Dunbar (1996), p. 174.

p. 147 Social ostracism: Eisenberger, N. I., et al. (2003), "Does rejection hurt? An fMRI study of social exclusion," *Science* 302, pp. 290–92.

p. 147 Ostracism by computer: Zadro, L., et al. (2004), "How low can you go? Ostracism by a computer lowers belonging, control, self-esteem, and meaningful existence," *J. of Experimental Social Psychology* 40, pp. 560–67.

p. 150 How gossip functions: Dunbar, R. (2004), "Gossip in evolutionary perspective," *Rev. of Gen. Psych.* 8 (2), pp. 100–10; Kurland, N., and Pelled, L. H. (2000), "Passing the word: Toward a model of gossip and power in the workplace," *Academy of Management Review* 25 (2), pp. 428–38.

p. 151 Grapevine accuracy: Davis, K. (1973), "The care and cultivation of the corporate grapevine," *Dun's Review* 102, pp. 44–47; Daft, R., and Steers, R. M. (1986), *Organizations: A Micro/Macro Approach,* Glenview, IL: Scott, Foresman; Smith, B. (1996), "Care and feeding of the grapevine," *Management Review* 85 (2), p. 6.

p. 152 Reviewing the moral dossier: Boehm, C. (1999), *Hierarchy in the Forest: The Evolution of Egalitarian Behavior,* Cambridge: Harvard University Press, p. 73.

p. 152 Grasso's outlandish pay: *New York Times,* October 5, 2004.

9 BANG BANG, KISS KISS

p. 154 Two porcupines: *New York Times,* April 4, 2004, p. 21.
p. 154 Dewey Ballantine: *NY Law Journal,* January 28, 2004, p. 10; *New York Times,* February 7, 2004.
p. 156 Extreme honesty: Kraman, S. S., and Hamm, G. (1999), "Risk management: Extreme honesty may be the best policy." *Annals of Internal Medicine* 131 (12), pp. 963–67; see also *Wall Street Journal,* May 18, 2004.
p. 156 Johns Hopkins death: *Hopkins Medicine,* spring/summer 2004.
p. 157 The bishop apologizes: *Dallas Morning News,* July 11, 1998.
p. 158 Ignorance about human reconciliation: de Waal, F. (1999), *Peacemaking Among Primates,* Cambridge: Harvard University Press, p. 43.
p. 159 How humans reconcile: ibid., pp. 238–39.
p. 160 The frequency of reconciliation: Preuschoft, S., et al. (2002), "Reconciliation in captive chimpanzees: A reevaluation with controlled methods," *Intl. J. of Primatology* 23 (1), pp. 29–50.
p. 162 Microsoft and Sun make up: *Washington Post,* April 3, 2004, April 5, 2004; *New York Times,* April 4, 2004.
p. 163 Fidgety bystanders: Aureli, F., and de Waal, F. (eds.) (2000), *Natural Conflict Resolution,* Berkeley: University of California Press, pp. 206–7.
p. 163 Mama: de Waal, F. (1982), *Chimpanzee Politics,* New York: Harper and Row, p. 56; de Waal (1999), pp. 2–22.
p. 163 Peacemaking wives: http://news.com.com/2008-1014-5184372.html?tag=nl.
p. 164 USS *Greenville* incident: *Honolulu Star-Bulletin,* March 12, 2001; O'Hara, E. A., and Yarn, D. (2002), "On apology and consilience," *Washington Law Rev.* 77, pp. 1122–92.
p. 164 Reconciliation outside the will of the victim: O'Hara and Yarn (2002).
p. 165 Likelier to reconcile with social allies: Aureli and de Waal (2000), p. 117.
p. 166 The legal risks of apology: Wagatsuma, H., and Rosett, A. (1986), "The implications of apology: Law and culture in Japan and the United States," *Law and Society Rev.* 20 (4), pp. 499–507.
p. 166 Tort law encourages only least therapeutic apologies: Shuman, D. W. (2000), "The role of apology in tort law," *Judicature* 180, pp. 180–89.
p. 167 A tender moment: Tavuchis, N. (1991), *Mea Culpa: A Sociology of Apology and Reconciliation,* Stanford: Stanford University Press, pp. 88–89.
p. 167 Bad Harvey: Biskind, P. (2004), *Down and Dirty Pictures: Miramax, Sundance, and the Rise of Independent Film,* New York: Simon and Schuster, p. 69.
p. 167 Voluntary versus forced reconciliation: Aureli and de Waal (2000), p. 46.
p. 168 Rhesus learn to become peacemakers: ibid., p. 117.
p. 170 Acting like gods or robots: Tavuchis (1991), p. 149.
p. 170 Moralistic aggression: O'Hara and Yarn (2002), p. 1153.
p. 170 Apology as exploitation: ibid., pp. 1186–87.

p. 171 Toro: Cohen, J. R. (1999), "Apology and Organizations: Exploring an example from medical practice," *Fordham Urb. L. J.* VII, pp. 1447–82.

p. 174 Cleanerfish: Bshary, R., and Würth, M. (2001), "Cleaner fish *Labroides dimidiatus* manipulate client reef fish by providing tactile stimulation," *Proceedings of the Royal Society of London, Series B—Biological Sciences* 268, pp. 1495–501.

10 MAKING FACES

p. 175 An insulted monkey: Darwin, C. (1872 [1998]), *The Expression of the Emotions in Man and Animals,* New York: Oxford University Press, p. 144.

p. 176 Darwin's dog: ibid., p. 62.

p. 177 Rat-faced: Gaufo, G. O., et al. (2003), "Hox3 genes coordinate mechanisms of genetic suppression and activation in the generation of branchial and somatic motor neurons," *Development* 130 (21), pp. 5191–201.

p. 177 How expressions evolved: Preuschoft, S. (2000), "Primate faces and facial expressions," *Social Research* 67, pp. 245–71.

p. 177 CIA polygraphers: Ekman, P. (2001), *Telling Lies,* New York: Norton, p. 285.

p. 177 Medical professionals: Kappesser, J., and Williams, A. C. (2002), "Pain and negative emotions in the face: judgements by health care professionals," *Pain* 99 (1–2), pp. 197–206.

p. 179 Facial action coding: Ekman, P., and Friesen, W. V. (1975), *Unmasking the Face,* Englewood Cliffs, NJ: Prentice Hall.

p. 182 Multiple human smile types: Ekman (2001), pp. 149–61.

p. 183 Lie detection by stroke victims: Etcoff, N., et al. (2000), "Lie detection and language comprehension," *Nature* 405, p. 139.

p. 185 Microexpressions: Ekman (2001), pp. 129–33.

p. 188 Exploiting facial expressions: Hill, D. (2003), *Body of Truth,* New York: Wiley.

p. 190 Samurai and the avoidance of expressions: Hall, E. (1976), *Beyond Culture,* Garden City, NY: Anchor, pp. 57–58.

p. 191 The talons of social awkwardness: Hall, J. A., and Halberstadt, A. G. (1986), "Smiling and gazing," in Hyde, J. S., and Inn, M. C. (eds.), *The Psychology of Gender,* Baltimore: Johns Hopkins University Press.

p. 191 Smiling: LaFrance, M., et al. (2003), "The contingent smile: A meta-analysis of sex differences in smiling," *Psychological Bulletin* 129 (2), pp. 305–34.

p. 192 Shoot-to-kill facial expressions: *New Yorker,* August 5, 2002.

p. 193 Poindexter: Ekman (2001), pp. 293–97.

p. 195 Facial makeup: Preuschoft (2000).

p. 195 Lip smack and air kiss: Schmidt, K. L., and Cohn, J. F. (2001), "Human facial expressions as adaptations: Evolutionary questions in facial expression research," *Yearbook of Physical Anthropology* 44, pp. 3–24.

11 FACIAL PREDESTINATION

p. 196 Everyone sees what you appear to be: Machiavelli, N. (1515 [2004]), *The Prince,* translated by Marriott, W. K., at http://etext.library.adelaide.edu.au/m/machiavelli/niccolo/m149p/chapter18.html.

p. 196 Aristotle and eye size: Sassi, M. M. (2001), *The Science of Man in Ancient Greece,* Chicago: University of Chicago Press.

p. 196 Darwin's nose: Zebrowitz, L. A. (1997), *Reading Faces: Window to the Soul?* Boulder, CO: Westview Press, p. 1.

p. 197 West Point: Mueller, U., and Mazur, A. (1996), "Facial dominance of West Point cadets as a predictor of later military rank," *Social Forces* 74, pp. 823–50.

p. 198 Baby-faced in Boston: Zebrowitz (1997), pp. 112–13.

p. 198 The face guy in the executive suite: ibid., pp. 101–2.

p. 199 Executive hair: Lutz, R. A. (2003), *Guts,* New York: Wiley, p. 180.

p. 200 Key stimuli: Zebrowitz (1997), pp. 68–69.

p. 201 Coloration in baboons: ibid., p. 70.

p. 202 Newborns like 'em pretty: Slater, A., et al. (1998), "Newborn infants prefer attractive faces," *Infant Behavior and Development* 21, pp. 345–54.

p. 203 It's good to be symmetrical: Grammer, K., and Thornhill, R. (1994), "Human *(Homo sapiens)* facial attractiveness and sexual selection: The role of symmetry and averageness," *J. of Comparative Psychology* 108, pp. 233–42.

p. 204 Looking masculine or feminine: www.brandeis.edu/gsa/gradjournal/2004/v.rennenkampff2004.pdf.

p. 205 Sir Isaac Newton's game face: Fara, P. (2003), "Face values: How portraits win friends and influence people," *Science* 229 (5608), pp. 831–32.

p. 205 Fiorina on the diminutive male brain: Anders, G. (2003), *Perfect Enough,* New York: Portfolio, p. 50.

12 MONKEY SEE...

p. 209 Fashion: Thoreau, H. D. (1854 [1986]), *Walden,* New York: Penguin, p. 32.

p. 209 Shirt color at IBM: Gerstner, L. (2002), *Who Says Elephants Can't Dance?,* New York: HarperBusiness, pp. 21–22.

p. 209 Chickens and ants: Hatfield et al. (1994), *Emotional Contagion,* Cambridge: Cambridge University Press, pp. 45–46.

p. 210 Routine vocal mimicry: ibid., p. 28.

p. 211 Chuck Yeager's accent: Wolfe, T. (1979), *The Right Stuff,* New York: Farrar Straus and Giroux, pp. 34–35.

p. 211 Phantom traffic jams: Strogatz, S. H. (2003), *Sync: The Emerging Science of Spontaneous Order,* New York: Hyperion, pp. 269–71.

p. 212 The selfish herd: Hamilton, W. D. (1971), "Geometry for the selfish herd," *J. Theor. Biol.* 31, pp. 295–311.

p. 213 Monkey face-reading: Preuschoft, S. (2000), "Primate faces and facial expressions," *Social Research,* 67, pp. 245–71.

p. 215 Synchrony through conversation: Hall, E. (1976), *Beyond Culture,* Garden City, NY: Anchor, p. 68.

p. 216 Using synchrony to judge social liking: Grahe, J. E., and Bernieri, F. J. (2002), "Self-awareness of judgment policies of rapport," *Personality and Social Psychology Bulletin* 28 (10), pp. 1407–18.

p. 216 Smile and be happy: Ekman, P. (2003), *Emotions Revealed,* New York: Holt, p. 36.

p. 216 Smile and be cool-headed: McIntosh, D. N., et al. (1997), "Facial movement, breathing, temperature, and affect: implications of the vascular theory of emotional efference," *Cognition and Emotion* 11 (17), pp. 171–95.

p. 217 Emotional contagion at the dentist's: Hatfield et al. (1994), pp. 193–94.

p. 217 Contagious enthusiasm: Ward, G. (1992), *A First-Class Temperament: The Emergence of Franklin Roosevelt,* New York: HarperCollins, pp. 221–22.

p. 218 Emotions across the counter: Pugh, S. D. (2002), "Service with a smile: Emotional contagion in the service encounter," *Academy of Management J.* 44 (5), pp. 1018–27.

p. 219 People as walking mood inductors: Barsade, S. G. (2002), "The ripple effect: Emotional contagion and its influence on group behavior." *Administrative Science Quarterly* 47, pp. 644–75.

p. 220 Mimicry for money: Van Baaren, R. B., et al. (2003), "Mimicry for money: Behavioral consequences of imitation," *J. of Experimental Social Psychology* 39, pp. 393–98.

p. 221 Stuttering: Hatfield et al. (1994), pp. 22, 33.

p. 221 Imitating your way to success: ibid., p. 176.

p. 221 Hazards of upward mimicry: Morris, D. (1979), *Manwatching,* New York: Abrams.

p. 222 The skipper: Poole, R. (2004), *Explorers House: National Geographic and the World It Made,* New York: Penguin.

p. 222 Matsushita as *maneshita: Forbes,* February 2, 2004.

p. 222 SpinBrush sales: *Business Week,* August 12, 2002.

p. 222 Protecting Premarin: *Wall Street Journal,* September 9, 2004.

p. 223 Copying Jack Welch's contract: *Chicago Tribune,* October 27, 2002.

p. 223 Imitating management fads: Strang, D., and Macy, M. W. (2001), "In Search of Excellence: Fads, Success Stories, and Adaptive Emulation," *Am. J. of Sociology* 107 (1), pp. 147–82.

p. 227 Boids: Reynolds, C. W. (1987), "Flocks, herds, and schools: A distributed behavior model," *Computer Graphics* 21 (4), pp. 25–34, http://www.red3d.com/cwr/boids/.

p. 231 Linux: http://www.li.org/linuxhistory.php.

p. 232 Linux challenges Microsoft: *Harvard Business Review,* September 2004.

p. 232 Self-organization at Harrah's, BP: ibid.

p. 233 Chimps at the water cooler: Yerkes, R. M. (1943), *Chimpanzees: A Laboratory Colony,* New Haven: Yale University Press, p. 52.

13 BUNNIES FOR LUNCH

p. 235 Predatory publisher: *New York Times Magazine,* July 20, 2003.
p. 236 Eight-hundred-pound gorillas: *Observer* (UK), August 31, 2003.
p. 237 Takeover piranha: *Wall Street Journal,* January 24, 2001.
p. 238 Shaving time: *New York Times,* April 4, 2004.
p. 239 Piggybacking on national chains: *Contra Costa* (CA) *Times,* November 11, 2004.
p. 239 Applying economic theory to animal choices: Stephens, D. W., and Krebs, J. R. (1986), *Foraging Theory,* Princeton: Princeton University Press.
p. 243 Focusing on the wrong prey at Scott Paper: Dunlap, A. (1996), *Mean Business,* New York: Fireside, pp. 139–42.
p. 243 Salespeople chasing the wrong accounts: *McKinsey Quarterly,* 2004 (3).
p. 243 Grizzly bear feeding habits: Conniff, R. (1998), *Every Creeping Thing,* New York: Holt.
p. 244 Teaching IBM to survive in the jungle: Gerstner, L. (2002), *Who Says Elephants Can't Dance?,* New York: HarperBusiness, pp. 176–77.
p. 245 Polaroid: *CFO Magazine,* January 1, 2003.
p. 245 Two hundred thousand unwanted cars: *Detroit News,* February 29, 2004.
p. 245 $5,000 rebates: *Crain's Cleveland Business,* May 24, 2004; *Business Week,* May 31, 2004.
p. 245 Toyota's flexibility: *Detroit News,* February 29, 2004.
p. 246 Internet statistics: http://www.sims.berkeley.edu/research/projects/how-much-info-2003/internet.htm. See also http://www.isoc.org/guest/zakon/Internet/History/HIT.html.
p. 249 Tree frog: Warkentin, K. M. (1995). "Adaptive plasticity in hatching age: A response to predation risk trade-offs," *Proceedings of the National Academy of Sciences* 92, pp. 3507–10.

14 A LANDSCAPE OF FEAR

p. 254 The advantages of acting like an unpredictable jerk: Silk, J. B. (2002), "Practice random acts of aggression and senseless acts of intimidation: the logic of status contests in social groups," *Evolutionary Anthropology* 11, pp. 221–25.
p. 255 Doing the jerk's bidding without question, Aureli, F., and de Waal, F. (eds.) (2000), *Natural Conflict Resolution,* Berkeley: University of California Press, p. 205.
p. 255 Claparède's pinprick: LeDoux, J. (1996), *The Emotional Brain: The Mysterious Underpinnings of Emotional Life,* New York: Simon and Schuster, p. 180; Claparède, E., in Rapaport D. (ed.) (1951), *Organization and Pathology of Thought,* New York: Columbia University Press.
p. 256 Cats with no cerebral cortex: LeDoux (1996), p. 79.
p. 257 Parmalat: *Wall Street Journal,* December 26, 2003.
p. 257 Redirected aggression: Virgin, C. E., and Sapolsky, R. (1997), "Styles of

male social behavior and their endocrine correlates among low-ranking baboons," *Am. J. of Primatology* 42, pp. 25–39.

p. 258 Avoiding ulcers by giving them: Sapolsky, R., in Aureli and de Waal (2000), pp. 114–16.

p. 259 Crushing the enemy: D'Aveni, R. (1995), *Hypercompetition,* New York: Free Press, p. 377.

p. 259 Lemon-lime "Overlord": Enrico, R., and Kornbluth, J. (1986), *The Other Guy Blinked,* New York: Bantam, p. 66.

p. 260 Retail is war: *Forbes,* March 5, 2001; http://www.limitedbrands.com/about/ltd/index.jsp.

p. 261 Ignoring anger at Miramax: Biskind, P. (2004), *Down and Dirty Pictures: Miramax, Sundance, and the Rise of Independent Film,* New York: Simon and Schuster, p. 74.

p. 261 Bullying percentages: http://www2.newpaltz.edu/~neumanj/neuman_and_keashly_siop2003.pdf.

p. 262 People who have a bullying boss: *San Francisco Chronicle,* October 19, 1998; Keashly, L., and Jagatic, K. (2000), "The Nature, Extent, and Impact of Emotional Abuse in the Workplace: Results of a Statewide Survey," presented at an Academy of Management seminar, August 4–9.

p. 263 Predators keep guppies sharp: O'Steen, S., et al. (2002), "Rapid evolution of escape ability in Trinidadian guppies *(Poecilia reticulata),*" *Evolution* 56 (4), pp. 777–84.

p. 264 Nurturing HP's softer side: Anders, G. (2003), *Perfect Enough,* New York: Portfolio, pp. 36, 40.

p. 264 High Machs: Wilson, D. S. (1998), in Dugatkin, L. E., and Reeve, H. K. (eds.), *Game Theory and Animal Behavior,* New York: Oxford University Press, p. 266.

p. 265 The allure of the mean friend: Hawley, P. H. (2003), "Prosocial and coercive configurations of resource control in early adolescence: A case for the well-adapted Machiavellian," *Merrill-Palmer Quarterly* 49, pp. 279–309; Hawley, P. H., et al. (manuscript under review), "The allure of the mean friend: Relationship quality and processes of aggressive adolescents."

p. 267 Durk Jager at P&G: *Wall Street Journal,* December 11, 1998, August 31, 2000.

p. 267 Mack the Knife: *Wall Street Journal,* June 28, 2004.

p. 267 Kumar and Computer Associates: *Wall Street Journal,* September 23, 2004.

p. 268 NASA: Columbia Accident Investigation Board, *Newsday,* August 27, 2003.

p. 268 The $20 million layoff rage: *New York Times,* August 1, 2001.

p. 270 A change of culture for baboons: Sapolsky, R. M., and Share, L. J. (2004), "A pacific culture among wild baboons: its emergence and transmission," *Public Library of Science* 2, p. 106, at http://www.plosbiology.org.

p. 272 Redwings keep it low-key: Hansen, A. J., and Rohwer, S. (1986), "Coverable badges and resource defence in birds," *Anim. Behav.* 34, pp. 69–76.

15 RUNNING WITH THE PACK

p. 273 Intel as an ant colony: *Sunday Oregonian* (Portland), January 13, 2002.
p. 273 Sheep faces: *New York Times,* September 14, 2004.
p. 274 MCI: *Wall Street Journal,* July 13, 2004.
p. 275 Sacrificing for the small group: Nicholson, N. (2000), *Executive Instinct: Managing the Human Animal in the Information Age,* New York: Crown, p. 31.
p. 276 A company of heroes: Ambrose, S. E. (1992), *Band of Brothers,* New York: Simon and Schuster, p. 307.
p. 276 NYNEX: Euchner, J., and Sachs, P. (1993), "The benefits of intentional tension," *Communication of the ACM* 36 (4).
p. 277 The trolley dilemma: Greene, J. D., et al. (2001), "An fMRI investigation of emotional engagement in moral judgment," *Science* 293, pp. 2105–8.
p. 279 Group size: Hackman, J. R. (2002), *Leading Teams: Setting the Stage for Great Performance,* Boston: Harvard Business School Press, p. 119.
p. 279 Group relationships, hearing distance: Dunbar, R. (1996), *Grooming, Gossip, and the Evolution of Language,* London: Faber, pp. 64, 122.
p. 280 Avoiding serious fights on first encounter: Mendoza, S. P. (1993), "Social Conflict on First Encounters," in Mason, W. A., and Mendoza, S. P. (eds.), *Primate Social Conflict,* New York: New York University Press, p. 85.
p. 281 Virtual work groups: Majchrzak, A., et al. (2004), "Can absence make a team grow stronger?" *Harvard Business Review,* May, pp. 131–37.
p. 281 The managing director as a dominatrix: Rolfe, J., and Troob, P. (2000), *Monkey Business,* New York: Warner, p. 220.
p. 283 If your bottler drives a Cadillac: *Fortune,* May 31, 2004.
p. 283 The user-exit subroutine: Ullman, E. (1997), *Close to the Machine,* San Francisco: City Lights Books.
p. 284 *Buttfuck* as a business term: Swartz, M., and Watkins, S. (2003), *Power Failure: The Inside Story of the Collapse of Enron,* New York: Doubleday.
p. 284 Getting invited to the joke: Cruver, B. (2002), *Anatomy of Greed: The Unshredded Truth from an Enron Insider,* New York: Avalon, p. 37.
p. 284 Inheritance in monkeys: Cheney, D. L., and Seyfarth, R. M. (1990), *How Monkeys See the World,* Chicago: University of Chicago Press, pp. 29–33; Silk, J. B., "The evolution of cooperation in primate groups," in Gintis, S., et al. (2005), *Moral Sentiments and Material Interests: The Foundations of Cooperation in Economic Life,* Cambridge: MIT Press.
p. 285 Family companies prosper: Anderson, R., and Reeb, D. M. (2003), "Founding family ownerships and firm performance: Evidence from the S&P 500," *J. of Finance* 58, pp. 1301–29.
p. 286 But they fall apart after the founder: http://knowledge.wharton.upenn.edu/papers/1284.pdf.
p. 286 Dogs in narcotics detection: Helton, W. S. (2004), "The development of expertise: Animal models?" *J. of General Psych.* 131 (1), pp. 86–96.
p. 289 Whales: Whitehead, H., and Rendell, L. (2004), "Movements, habitat use

and feeding success of cultural clans of South Pacific sperm whales," *J. of Animal Ecology* 73, pp. 190–96.

p. 289 Cowbirds: Smith V., et al. (2000), "A role of her own: Female cowbirds, *Molothrus ater,* influence the development and outcome of song learning," *Animal Behav.* 60, pp. 599–609.

p. 289 Sexy sparrows: MacDougall-Shackleton, E. A., and MacDougall-Shackleton, S. A. (2001), "Cultural and genetic evolution in mountain white-crowned sparrows," *Evolution* 55, pp. 2568–75.

p. 289 Silicon Valley savages: English-Lueck, J. A. (2002), *Cultures@SiliconValley.* Stanford: Stanford University Press, p. 24.

p. 290 Culture hides itself: Hall, E. T. (1959), *The Silent Language,* Garden City, New York: Doubleday, p. 53.

p. 291 Missing mission statements: Wurman, R. (2001), *Information Anxiety 2,* Indianapolis: Que, p. 132.

p. 291 Wal-Mart's unwieldy culture: *Harvard Business Review,* July-August, 2004, p. 37.

p. 292 Unhappy at American: *New York Times,* April 27, 2004.

p. 293 Workers like convicts going to jail: Gallagher, R. S. (2003), *The Soul of an Organization,* Chicago: Dearborn, pp. 8–9.

p. 294 Badges of status: D'Aveni, R. A., and Kesner, I. F. (1993), "Top managerial prestige, power and tender offer response: A study of elite social networks and target firm cooperation during takeovers," *Organization Science* 4 (2), pp. 123–51.

p. 294 GM-EDS divorce: Levin, D. (1989), *Irreconcilable Differences,* Boston: Little, Brown, p. 300.

p. 295 Stealing from IBM grandkids: Gerstner, L. (2002), *Who Says Elephants Can't Dance?* New York: HarperBusiness, p. 204.

p. 296 In-groups for beating up out-groups: Wrangham, R., and Peterson, D. (1996), *Demonic Males,* New York: Houghton Mifflin, pp. 196–97.

p. 297 Encapsulation at GM: Levin (1989), p. 340.

p. 298 East African in- and out-groups: Brewer, M. B. (1999), "The psychology of prejudice: In-group love or out-group hate?" *J. of Social Issues,* 55 (3), pp. 492–544.

p. 299 Wolf behavior: Steinhart, P. (1995), *The Company of Wolves,* New York: Vintage.

EPILOGUE

p. 302 The moose on the table: *Wall Street Journal,* February 4, 2003.

p. 303 Chimpanzees and us: Ridley, M. (1999), *Genome,* New York: HarperCollins, p. 29.

p. 306 Benevolent alphas: Taylor, S. E. (2002), *The Tending Instinct,* New York: Times Books, p. 156.

p. 307 Cocker spaniels: *Newsweek,* September 30, 2002.

ACKNOWLEDGMENTS

This book would not have been possible without the contributions of many kind informants in the corporate workplace who must remain nameless in the interest of continued gainful employment. My thanks to you all. Thanks also to my agent, John Thornton, who suggested the topic for this book, and to editors Emily Loose and John Mahaney at Crown. Thanks to my many editorial friends at *Smithsonian Magazine,* particularly Jim Doherty, Carey Winfrey, Beth Py-Lieberman, Helen Starkweather, and Sally Maran; at *National Geographic,* thanks to Bob Poole, Jennifer Reek, and Kathy Maher; at *Discover Magazine,* thanks to Steve Petranek and Dave Grogan. In television, I am especially grateful to Bonnie Benjamin-Phariss of Vulcan Productions, who got me thinking (against my instincts) about the affiliation instinct. Thank you also to Jason Williams of JWM Productions, Chris Weber of Tiger/Tigress, and Mick Kaczorowski of the Discovery Channel. Many friends and strangers gave advice along the way, including Al Vogl, Ben Haimowitz of the Academy of Management, Baba Marietta of Michigan State University, Melissa Gerald, Amy Solomon, Dan Merchant, Laura Betzig, Gene Murphy, and Deby Cassill. Thank you finally to a snarling, baying pack of Conniffs, including James C. G., Karen, Greg, Jamie, Ben, and Clare, for your help, your patience, and your occasional derision.

INDEX

ABOUT THE AUTHOR

RICHARD CONNIFF, the author of *The Natural History of the Rich,* is an award-winning journalist whose work has appeared in *Smithsonian, Atlantic Monthly, The New York Times Magazine,* and *National Geographic.*

ABOUT THE AUTHOR

RICHARD CONNIFF, the author of *The Natural History of the Rich,* is an award-winning journalist whose work has appeared in *Smithsonian, Atlantic Monthly, The New York Times Magazine,* and *National Geographic.*